Laboratory Manual for Tobin and Duscheck's
ASKING ABOUT LIFE THIRD EDITION

Marni Fylling

THOMSON
™
BROOKS/COLE

Australia • Canada • Mexico • Singapore • Spain • United Kingdom • United States

THOMSON

✳ ™

BROOKS/COLE

Executive Editor, Life Sciences: *Nedah Rose*
Editor-in-Chief: *Michelle Julet*
Development Editor: *Anne Scanlan-Rohrer*
Assistant Editor: *Kari Hopperstead*
Editorial Assistant: *Sarah Lowe*
Technology Project Manager: *Travis Metz*
Marketing Manager: *Ann Caven*
Marketing Assistant: *Leyla Jowza*
Advertising Project Manager: *Kelley McAllister*
Project Manager, Editorial Production: *Kelsey McGee*

Print/Media Buyer: *Judy Inouye*
Permissions Editor: *Joohee Lee*
Production Editor: *Denise DeLancey*
Text Designer: *Jeanne Calabrese*
Photo Researcher: *Kathleen Olson*
Cover Designer: *Larry Didona*
Cover Image: *Tui de Roy/Minden Pictures*
Illustration: *Marni Fylling*
Compositor and Production Service: *Graphic World, Inc.*
Printer, Cover Printer and Binder: *Quebecor World–Dubuque*

Printed in the United States of America
1 2 3 4 5 6 7 07 06 05 04

For more information about our products, contact us at:
Thomson Learning Academic Resource Center
1-800-423-0563
For permission to use material from this text, contact us by:
Web: http://www.thomsonrights.com

Library of Congress Control Number: 2004101707

ISBN 0-534-40659-9

Brooks/Cole—Thomson Learning
10 Davis Drive
Belmont, CA 94002
USA

Asia
Thomson Learning
5 Shenton Way #01-01
UIC Building
Singapore 068808

Australia/New Zealand
Thomson Learning
102 Dodds Street
Southbank, Victoria 3006
Australia

Canada
Nelson
1120 Birchmount Road
Toronto, Ontario M1K 5G4
Canada

Europe/Middle East/Africa
Thomson Learning
High Holborn House
50/51 Bedford Row
London WC1R 4LR
United Kingdom

Latin America
Thomson Learning
Seneca, 53
Colonia Polanco
11560 Mexico D.F.
Mexico

Spain/Portugal
Paraninfo
Calle Magallanes, 25
28015 Madrid, Spain

Contents

Preface

This laboratory manual is designed to accompany the third edition of *Asking About Life* by Jennie Dusheck and Allan Tobin. Lab topics from six units of the text are covered, and each of the 15 labs is identified with the corresponding chapter or chapters from *Asking About Life*. Although consistent with the definitions and scope of *Asking About Life*, this manual may be used with any introductory biology textbook.

Biology, like any other academic subject, requires the memorization of terminology and of details of processes. However, the real aims are to teach you how to ask questions and to give you the equipment to find answers. Every day, you are faced with countless bits of information, from what you see outside to television news. A good biology course should increase your sense of wonder, interest, and understanding of the world, as well as your ability to form opinions and make decisions about information. Instead of merely accepting the results of a study in the newspaper, an informed reader asks how the study was conducted. Was the number of subjects used sufficient to eliminate individual variation? Does the graph accurately reflect the results, or are the results exaggerated?

The biology lab is the perfect place to teach such critical thinking skills. Biology is inherently fascinating, and labs should be fun and interesting. This lab manual employs activities that encourage involvement. You will often need to get up from your desk and interact with other students. Questions are designed to make you think instead of to merely make you repeat facts. As you approach these labs, it is less important to look at an unfamiliar animal and remember anatomical terms than to determine the function and value of that animal's adaptations by examining its anatomy and environment. Such critical thinking skills may be used in all aspects of life, not just in the circumstances presented in class, from interpreting information in the media to making decisions about health and the environment.

One goal of this lab manual is to use easily obtained materials; besides being practical for the often-limited budgets of schools, the use of familiar materials helps reinforce the concepts. For example, Lab 5, on enzymes, uses liver and hydrogen peroxide to illustrate the action and importance of enzymes. In addition to reinforcing the idea that enzymes exist in all cells (not just in mysterious solutions in test tubes), the use of common materials increases the chances that the materials will be encountered again. The liver at the grocery store or the hydrogen peroxide in the medicine cabinet will likely remind you of the tubes of bubbling liver puree from the enzyme lab.

Before each lab, read the lab exercise (and any related material from *Asking About Life*) so that the lab period is not spent reading. Come to each lab with a basic understanding of the concepts and activities.

Data sheets follow each lab. These sheets include space to draw, space to answer questions asked within and at the end of each lab, and tables and graphs to complete during experiments. At the beginning of each lab, tear out the data sheets; staple the data sheets together; and complete them as you progress through the lab. Data sheets will be turned in after the lab period and will be used to evaluate your performance.

The drawings you are asked to do within the labs are intended to help you better understand the lab material. Drawing a specimen requires you to examine the specimen carefully and to locate the structures referred to in the text. Your drawings need not be beautiful; what is important is that they correctly identify the specimen and the location of important structures. A beautifully executed drawing that incorrectly represents the specimen is useless.

After every lab period, clean your area and replace all lab materials as instructed.

I hope that these labs encourage your interest in and understanding of the world around you. Please contact me at marni_fylling@hotmail.com with comments or suggestions about this lab manual.

To the Instructor

A preparation guide for these labs is available from the Instructor Book Companion site at http://www.biology.brookscole.com.

Acknowledgments

The development of this manual was helped immeasurably by the careful review of lab instructors around the country: Patricia Cox, University of Tennessee; Sara Brenizer, Shelton State Community College; and Diane Auer Jones, Community College of Baltimore, Catonsville Campus. Special thanks to Betsy Harris, Appalachian State University, and Gail Stratton, University of Mississippi. Eric Rabitoy from Citrus College provided constructive, thoughtful reviews, as well as ideas for labs on cellular respiration, photosynthesis, and reproduction/development. Thank you.

Thanks to the authors of *Asking About Life,* Jennie Dusheck and Allan Tobin, for their encouragement during the creation of this manual.

Many thanks to Amy Dunleavy, Susan Shipe, Charlene Squibb, Carol Bleistine, and especially Lee Marcott, who helped with all aspects of the first version of the lab manual. The combined efforts of all made production relatively painless. Thanks also to Denise DeLancey and Kari Hopperstead, who made working on the revision a pleasure.

Many friends and colleagues were always there to answer strange questions and to help solve problems; without their support, I would have felt alone. Many, many thanks to Stian Alesandrini, Martha Balogh, Mick Ellison, Jon Geller, Carole Kelley, Denise Lim, Rob Longair, Carl Mehling, Brian Ort, Lisa Soon, Dougald Scott, Kendell Silveira, John Welch, and Dennis Wilson.

Marni Fylling

Safety Guidelines

Always use common sense in the laboratory. Some materials are potentially dangerous and should be treated with care.

1. Read the lab exercise before you come to lab so that you are familiar with the procedures you will perform.

2. Personal safety

   a. Never eat, drink, apply makeup/lip balm, or smoke in lab.

   b. Wear closed-toe shoes at all times. No sandals are allowed in lab.

   c. Tie back long hair and remove dangling jewelry.

   d. Do not wear clothing that has loose sleeves or that can catch on equipment. Do not wear shorts or short skirts in lab. Your legs should always be covered.

   e. Some chemicals may damage your clothes. Wear appropriate clothing, an apron, or a lab coat.

   f. Be aware of the location of eyewashes, first aid kits, safety showers, fire blankets, and exits. Note the location of the nearest phones, and note the appropriate telephone numbers to call in case of an emergency.

   g. Wear approved safety goggles, gloves, and/or a lab coat when using dangerous chemicals, working with live organisms such as bacteria, or heating chemicals.

   h. Report any accidents, injuries, or spills to the instructor immediately. Wash chemicals or microorganisms from your skin immediately and thoroughly.

   i. Do not perform unauthorized experiments or use equipment until you are instructed in its use.

3. Using heat sources

   a. Do not leave heat sources such as hot plates, Bunsen burners, and so on unattended.

   b. Handle hot glassware with the appropriate apparatus: never attempt to handle hot glassware with bare hands.

   c. Use caution when heating liquids. Add boiling chips **before** any liquid has begun heating.

   d. Use care when heating glassware. Inspect your glassware thoroughly before adding chemicals. If any cracks are observed, replace the glassware.

   e. Materials heated in a test tube, flask, beaker, and so on can react suddenly and violently. Therefore, never point the mouth of a container being heated or undergoing a chemical reaction toward yourself or anyone else.

4. Use and disposal of chemicals

   a. Do not smell or taste chemicals unless lab procedure specifically requests you to do so.

   b. Dispose of all materials as instructed: many lab materials cannot be thrown in the garbage or poured down the sink.

   c. Do not put unused chemicals back into containers unless you are instructed to do so.

   d. Use water, chemicals, and other materials sparingly.

5. Keep all liquids away from electrical cords.

6. Be particularly careful when handling razor blades, knives, and glassware. Broken glass should be swept with a broom into a dustpan (not with or into your hands) and disposed of into special glass waste containers (not into the waste bin).

7. At the end of lab

   a. Always clean your area and properly replace all materials and equipment. Wipe down your bench area, and dispose of any chemicals as instructed.

   b. Wash your hands thoroughly with soap and water.

Lab 1   Snails/Experimental Design
This lab accompanies Chapter 1 of *Asking About Life.*

Materials/Equipment

Large glass Petri dishes

Enamel trays

Snails (two to three per group)

Dissecting microscopes

Materials for designing an experiment

Objectives

1. Observe snail anatomy and behavior

2. Learn how to use a dissecting microscope

3. Apply the characteristics of life to a snail

4. Perform a simple experiment using the scientific method

5. Design two experiments using the scientific method

Introduction

People are naturally curious about the world around them and, from early childhood, ask questions about how things work. Scientists maintain this curiosity and answer their questions using the **scientific method,** a formal set of rules for forming questions and testing ideas. In this lab, you will use the scientific method (see Chapter 1 of text) to learn more about the common garden snail.

In brief, the scientific method involves the following steps:

Scientific Method
A. Observe
   1. Ask a question
B. Develop a hypothesis
   1. Consult prior knowledge
C. Design a controlled experiment
D. Collect data
E. Interpret data
   1. Consult prior knowledge
F. Draw a conclusion

All recognized organisms have a **scientific name** made up of two words, usually in Latin or Greek, that describe the organism. The first word is the **genus,** which is always uppercase and is usually a noun. The second word is the **species,** which is always lowercase and is usually an adjective. Both names are italicized or underlined. You may be familiar with the scientific name for human beings, *Homo sapiens. Homo* is human in Latin, and *sapiens* is wise.

The scientific name of the common garden snail is *Helix aspera. Helix* is coil or spiral (as in the shell); *aspera* is rough, possibly a reference to the texture of the snail's skin. The genus name can often be used alone (or abbreviated if it has been used previously), so you can call your snail *Helix* or *H. aspera.*

H. aspera was originally brought to America from Europe, where it was used in cooking. It has proliferated here, where it is a garden pest that feeds on live and decaying plant matter.

Snails are closely related to slugs, clams, squid, oysters, and octopuses. You will be working with live snails, so **treat them with care.**

- Handle snails gently, especially when picking them up. Try sliding them across the surface a bit instead of pulling directly upward: you don't want to pull them out of their shells!
- Do not put any substances directly on the snails.
- Do not poke the snails with anything sharp.

Observation

First, choose a snail for observation. Like all animals, snails need food, water, and protection from predators, extreme temperatures, and drying out. Remember these requirements as you examine your snail—they will help you understand some of its behavior.

1. Put the snail on a glass Petri dish. Examine the coiled shell.

    Q1. How many different functions does the shell provide for the snail? What are these functions? Consult the list of requirements just given for ideas.

2. If the snail has pulled into its shell, give it a few minutes to come out. If it is still uncooperative, try another snail.

3. Find the two pairs of tentacles. One pair carries the eyes; the other pair is tactile (sensitive to touch) and chemosensitive (sensitive to particles you would smell or taste).

    Q2. Can you tell which pair is which? How does your snail use each pair of tentacles?

4. The large surface upon which the snail travels is its **foot.** Turn the dish upside down to see the undersurface of the foot. The foot secretes a slimy mucus as the snail travels.

    Q3. How does the snail move?

    Q4. What functions does the mucus serve?

Using a Dissecting Microscope

Models of dissecting microscopes vary considerably, but all have two eyepieces and several magnifications (see Figure 1-1). Microscopes are expensive and delicate equipment. Always carry a microscope with both hands (see Figure 1-2). If any surface of the microscope gets wet, clean it immediately with **lens paper.**

1. Turn on the **light,** which may be built into the microscope or separate from the microscope, depending on model.

2. Look through the **eyepieces.** You may need to adjust the distance between the eyepieces to match your eyes. Gently pull the eyepieces apart or push them together (as you look through them) until you can see one image with both eyes open. If you wear eyeglasses, try using the microscope without your glasses.

3. Put your snail in its dish on the **stage** of the microscope. Turn the **focus knob** until the snail is in focus. You may need to move the dish to keep your snail in view. Use the **magnification control** to change the magnification and understand the range available. The eyepieces contain lenses that magnify the snail; as you turn

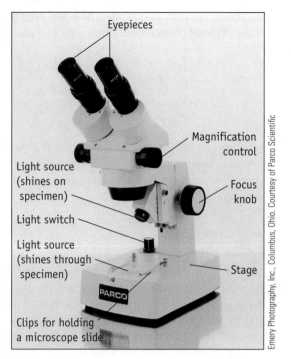

Figure 1-1
Parts of a Dissecting Microscope.

the magnification control, you change **objective lenses** inside the microscope to magnify the snail further.

4. Once you have your snail in focus, try changing the **light source.** Most dissecting microscopes have a light above the specimen that shines on the specimen and another light below that shines through the specimen.

5. If you have both types of lights, experiment with them.

   Q5. Which light gives you a better view of your snail?

   Q6. Can you see movement inside the snail?

   Do not leave your snail under the light too long because it will get too hot.

6. Draw your snail on the data sheet and label the structures you can identify.

Characteristics of Life

All living things

> Are organized into parts
>
> Perform chemical reactions
>
> Obtain energy from their surroundings
>
> Change with time
>
> Respond to the environment
>
> Reproduce
>
> Share a common evolutionary history
>
> *Q7. Is your snail alive? How do you know? What three observations have you made that prove your snail is alive?*

Figure 1-2
Microscope Use. Always hold a dissecting microscope with both hands.

*Q8. Has your snail shown behavior that exhibits **homeostasis,** the tendency to maintain a stable internal environment in response to a changing external environment?*

Experiment

For the first experiment, you will determine how a snail will respond to a vertical surface (see Figure 1-3). Will it travel upward, possibly looking for a way to escape? Will it travel downward, perhaps looking for shade, moisture, or food? Will it stay in one spot? Will it move sideways?

Develop a Hypothesis

To develop an experiment, a researcher must make careful observations and use the information gained to devise a **hypothesis,** an informed prediction about what will happen in different situations. The point of an experiment is to test the hypothesis to see if it is valid. Therefore, a hypothesis must be **testable:** it must be possible to devise an experiment that will disprove the hypothesis if it is incorrect.

Based on what you have observed so far, what do you suppose the snail will do when placed on a vertical surface? Write your hypothesis and the reasoning behind it on the data sheet.

Design an Experiment

To test your hypothesis, you will construct an **experiment.** Simply put a snail in the center of a dish or tray, hold the dish vertically, and wait for the snail to respond (if it does).

A good experiment always compares the results of the experiment with the results of a **control,** a version of the experiment in which all conditions are the same except the one factor (or **variable**) you are testing. It is then clear that differing results are a consequence of the variable, not other factors.

In this case, the variable is the vertical surface. Put a snail in the center of the same type of dish, but leave the dish on a horizontal surface.

Perform the Experiment

A. Trial 1

1. Get two snails and two clean dishes. (You may wash the dish you have been using.) Decide which snail is the experimental subject and which is the control, then put each snail in the center of its dish. Leave the control dish on the lab bench, but hold the experimental dish vertically, as shown in Figure 1-4.

2. Be patient. After your snails have had a chance to respond, write your results on the data sheet under "Trial 1."

B. Trial 2

1. A good experiment should be performed on several individuals to eliminate individual variation. Try this experiment one more time with two different snails. Make sure to clean and dry the dishes first: you do not want the slime trails from the first trial to affect the behavior of the snails in the second trial.

2. Record your results on the data sheet under "Trial 2."

    Q9. Did you get the same results in Trial 1 and Trial 2? Can you see why you would want to perform an experiment on many individuals before you decide that the data support your hypothesis?

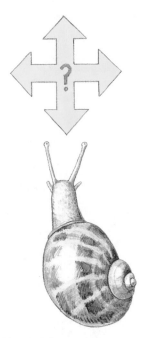

Figure 1-3
Which Direction Will Your Snail Travel?

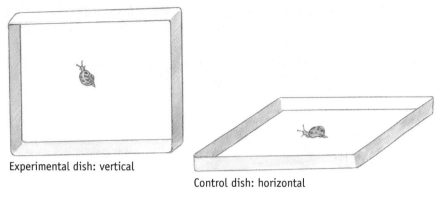

Experimental dish: vertical

Control dish: horizontal

Figure 1-4
Setup for Snail Experiment.

C. Scientists must communicate their results to get feedback from their peers, to help others with similar research, to educate, or to receive financial support for continued research. Compare your results with those of your classmates to see what they learned from performing the same experiment.

> *Q10. Are their results similar? Describe briefly how their results are similar or different from yours.*

D. Now, return to your original hypothesis. A good scientist is always willing to re-think his or her hypotheses when more data become available. When a hypothesis is revised, a new experiment may be performed to test the revised hypothesis.

> *Q11. Would the results of your experiment and the results obtained by other students cause you to modify your hypothesis? How would you modify your original hypothesis?*

Your Experiment

Now that you are getting to know your snail, you should be able to develop your own questions about snail behavior. There are several materials available to use to design your own experiments. You do not have to use all the materials, but be creative in how you use the materials you choose.

On the data sheet under "Your Experiment," write four questions that could be answered using these materials.

Your Hypotheses

Choose two of your questions and, based on what you have learned about snails so far, develop a hypothesis that answers each question. In other words, what do you suppose the snail would do when faced with the situation presented in your question? Write the hypotheses and the reasoning behind them on the data sheet.

Testing Your Hypotheses

Make sure that you design experiments that test only one variable; also, design a control for each experiment in which everything is the same except that variable. If you have time, try to repeat your experiments. Record your results on the data sheet.

> *Q12. Were your hypotheses supported or disproved by your results?*

> *Q13. Would you modify your hypotheses now that you have more information? How would you modify them?*

Q14. *Did you learn anything about experimental design? If you were going to test your hypotheses again, would you make changes to your experimental setup? What would those changes be?*

Q15. *Compare your results with those obtained by your classmates. Were the results similar? How do you account for differences?*

At the end of the lab

- Clean your area
- Put the materials, snails, and dissecting microscopes away properly
- Wash your hands

Snails
Data Sheet

Observation

Q1. *How many different functions does the shell provide for the snail? What are these functions? Consult the list under*

"Observation" at the beginning of the lab for ideas. _____

_____

_____

Q2. *Can you tell which pair of tentacles is which? How does your snail use the different tentacles?* _____

_____

_____

Q3. *How does the snail move?* _____

_____

_____

Q4. *What functions does the mucus serve?* _____

_____

_____

Using a Dissecting Microscope

Q5. *Which light gives you a better view of your snail?* _____

Q6. *Can you see movement inside the snail?* _____

_____

_____

Draw your snail on the data sheet and label the structures you can identify.

Characteristics of Life

Q7.   *Is your snail alive? How do you know? What three observations have you made that prove your snail is alive?*

_____

_____

_____

Q8.   *Has your snail shown behavior that exhibits homeostasis, the tendency to maintain a stable internal environment in response to a changing external environment?* _____

_____

_____

Experiment

Based on what you have observed so far, what do you suppose the snail will do when placed on a vertical surface?

Hypothesis: _____

Reasoning: _____

Trial 1

Experimental results: _____

Control results: _____

Trial 2

      Experimental results: _____

      Control results: _____

Q9. *Did you get the same results in Trial 1 and Trial 2? Can you see why you would want to perform an experiment on many individuals before you decide that the data support your hypothesis?* _____

_____

_____

Q10. *Are the results of your classmates similar to your results?* _____

_____

_____

Q11. *Would the results of your experiment and the results obtained by other students cause you to modify your hypothesis? How would you modify your original hypothesis?* _____

_____

_____

Your Experiment

Four questions that could be answered using material in the lab are

1. _____

2. _____

3. _____

4. _____

Your Hypotheses

      Hypothesis 1: _____

      Reasoning: _____

      Hypothesis 2: _____

      Reasoning: _____

Testing Your Hypotheses

Experiment 1
Hypothesis:

	TRIAL 1	TRIAL 2
Experimental results		
Control results		

Experiment 2
Hypothesis:

	TRIAL 1	TRIAL 2
Experimental results		
Control results		

Q12. *Were your hypotheses supported or disproved by your results?* _____

_____

Q13. *Would you modify your hypotheses now that you have more information?* _____

_____

_____

Q14. *Did you learn anything about experimental design? If you were going to test your hypotheses again, would you make*

changes to your experimental setup? What would those changes be? _____

_____

_____

Q15. *Compare your results with those obtained by your classmates. Were the results similar? How do you account for differences?* _____

_____

_____

Questions

Q16. *In your own words, define the following:*

Hypothesis: _____

Experimental variable: _____

Control: _____

Q17. *Why is it important to have a large number of trials before making conclusions about experiments?*

_____

_____

_____

Q18. *Why is it important to have many experimental subjects instead of one?* _____

_____

_____

Q19. *What is the importance of a control?* _____

_____

_____

Q20. *What is the value of comparing your results with the results of your peers?* _____

_____

_____

A scientist wants to determine the effects of a certain fertilizer on the growth of corn plants. She grows 80 plants in the same greenhouse and waters the plants with the same amount of fluid—water with different concentrations of fertilizer. Twenty plants are watered with plain water, 20 are watered with a weak solution of fertilizer, 20 are watered with a medium concentration of fertilizer, and 20 are watered with a high concentration of fertilizer. Over a month, she monitors the growth of the corn plants.

Q21. What was the control in this experiment? _____

Q22. How did the scientist make sure that any differences she measured were not the result of individual variation between

corn plants? _____

_____

_____

Q23. What were three ways the scientist ensured that growth conditions were the same for all plants? _____

_____

_____

An article in the newspaper describes a new heart medication that claims to greatly reduce the risk of heart disease. In the study, three patients took the medication for a year; none developed heart disease.

Q24. What are three things about this experiment that would make you skeptical of its conclusions? _____

_____

_____

Lab 2   Biochemistry
This lab accompanies Chapters 2 and 3 of *Asking About Life.*

HOMEWORK: Bring a food item for testing. Most juices, fruits, and vegetables, as well as bread or anything that mixes with water (flour, cornmeal, etc.), are appropriate.

Note: Foods that are oily, dense, and dry (such as uncooked beans, rice, or pasta); strongly colored (such as grape juice or beets); or highly acidic (such as oranges or grapefruit juice) do not work well in these experiments.

Materials/Equipment

2-inch squares (about 70) of kraft paper or brown paper grocery bag

Vegetable oil, 100 mL with 3 mL pipette

Sugar solutions, 100 mL each with 3 mL pipettes:

  5% glucose

  5% fructose

  5% sucrose

Lugol's iodine, three or four bottles with droppers

4% boiled starch solution, 100 mL with 3 mL pipette

0.5% $CuSO_4$, 100 mL with 3 mL pipette

10% NaOH, 150 mL with 3 mL pipette

Egg albumen solution

Razor blades

3 mL disposable, graduated pipettes

Boiling chips

Tub for used test tubes

Waste container for emptying test tubes

Wax pencils and label tape

Variety of foods for testing

Sharps disposal

Per group of three or four students:

  Hot plate

  Benedict's reagent, 20 mL with 1 mL pipette

  Deionized water, 20 mL with 1 mL pipette

  400 mL beaker for water bath

  20 test tubes

  Test tube racks

  Test tube holders

  Glass stir rod

Objectives

1. Understand the use of positive/negative controls

2. Learn the proper use of graduated pipettes

3. Recognize the four classes of biological molecules and test for the presence of these molecules in different food substances

Introduction

Chemistry is the branch of science that studies the properties and transformations of **matter** (anything that has mass and occupies space). All matter is made up of tiny units called **atoms.** Each atom consists of one or more **protons** (and sometimes **neutrons**) within a nucleus and one or more **electrons** that orbit around the nucleus. Atoms can combine with other atoms by forming bonds. In a **covalent bond,** atoms share electrons; in an **ionic bond,** electrons are transferred between atoms. A **molecule** is composed of two or more atoms in a chemical bond.

Biochemistry is the study of the chemistry of living organisms. All organisms consist of the same few kinds of small organic molecules, or **building blocks.** Table 3-2 in the text shows the 35 molecules that are the basis of nearly all organisms. These 35 molecules can form an extraordinary number of structures. Nearly all of these molecules are made of chains of carbon atoms, and many can link together to form larger molecules called **polymers.** Biological molecules are divided into the following four classes:

Lipids: fats, oils, waxes

Carbohydrates: sugars and the polymers formed by sugars, such as cellulose and starch

Amino acids: molecules that form the polypeptide chains that make up proteins

Nucleotides: molecules that form polymers called **nucleic acids,** a main component of RNA and DNA

Some organisms synthesize many of their own molecular building blocks. For instance, plants use carbon dioxide from the air; nitrogen, water, and minerals from the soil; and energy from sunlight to produce the lipids, carbohydrates, amino acids, and nucleotides they need to live. Most animals, including humans, can synthesize many of these molecules but rely on the food they consume to provide the rest.

Nucleotides have three parts: a sugar, one or more phosphate groups, and a nitrogenous base. Each of these parts has different functional groups, giving them several chemical properties. Nucleotides form the main structure of RNA and DNA. ATP, the energy source for all biological processes in all cells, is also a nucleotide. See "How Do Nucleotides Link Together to Form Nucleic Acids?" in the text for more about nucleotides.

It is fairly simple to test for three of the four classes of organic molecules and for the macromolecules they form in food samples. Each test you will perform uses an **indicator** to identify the presence of lipids, carbohydrates, or proteins. To recognize the effect of the indicator, you must perform two controls: a **positive control** and a **negative control.**

Positive and Negative Controls

A positive control uses an indicator with a sample known to have the substance for which you are testing. With this control, you will be certain that your test is accurate and that your indicator is working.

A negative control involves the use of an indicator with a sample that will not react with the indicator. In most cases, the negative control will be water, which contains none of the substances for which you are testing.

You will compare your experimental results with your control results to determine whether your sample contains the substance for which you are testing.

Using a Graduated Pipette

For most of the chemical tests, you will need to measure and dispense solutions with a graduated pipette (see Figure 2-1). To measure accurately and keep from spilling or splattering chemicals, use pipettes as follows:

1. While leaving the pipette in the container, raise the pipette tip above the level of the solution. Gently squeeze the pipette bulb to expel any solution it already contains.

2. With the pipette tip above the level of the solution, squeeze the bulb slightly, **THEN** lower the tip into the solution and **SLOWLY** release the bulb, a little at a

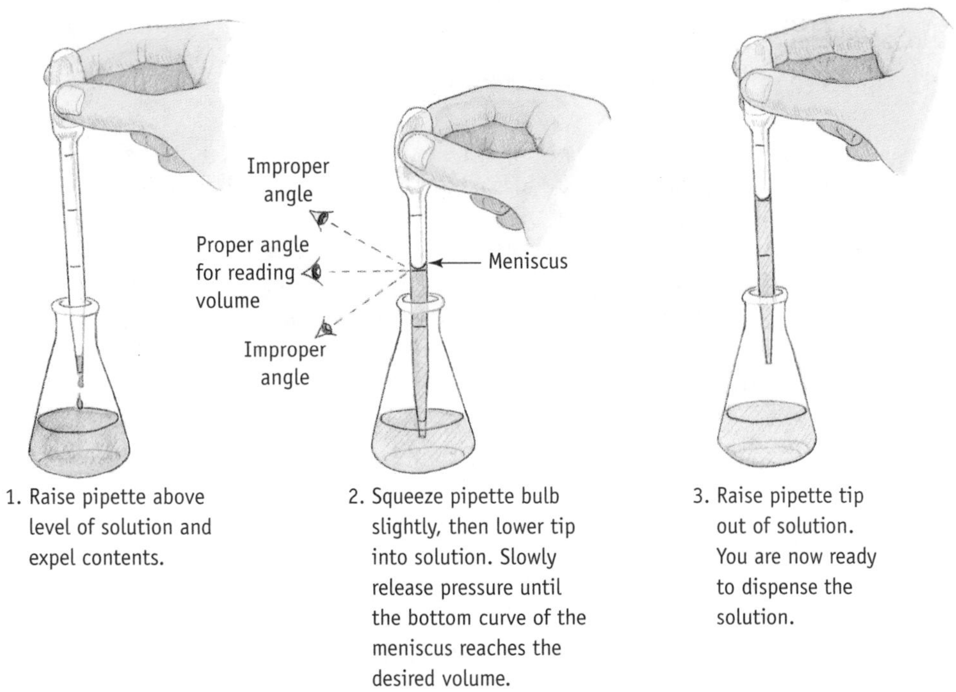

Figure 2-1
How to Use a Graduated
Pipette.

Improper angle

Proper angle for reading volume

Meniscus

Improper angle

1. Raise pipette above level of solution and expel contents.

2. Squeeze pipette bulb slightly, then lower tip into solution. Slowly release pressure until the bottom curve of the meniscus reaches the desired volume.

3. Raise pipette tip out of solution. You are now ready to dispense the solution.

time, watching the level of the solution in the pipette rise until it reaches the desired volume. To correctly read this volume, your eye should be directly in line with the curved top surface of the solution, the **meniscus.** The bottom of the meniscus marks the volume of solution in the pipette.

3. Raise the pipette tip out of the solution. At this point, you can release pressure on the bulb. You are now ready to dispense the solution.

Lipids

Lipids include oils, fats, and waxes. Lipids are the main component of the plasma membrane of all cells. In addition, they contain more chemical energy per unit weight than other biological molecules, so lipids are often used by organisms for energy storage. Humans store energy in the form of fat, as do many other animals and plants. Vegetable oils (sunflower, olive, corn, etc.) come from the stored fat in the seeds or fruits of those plants: this is energy intended to help a new plant grow. Lipids are also important to animals for insulation and to help the body store fat-soluble vitamins, such as vitamin E and vitamin A. Have you ever noticed how vitamins are packaged into pills? Vitamin A and vitamin E are in an oily substance (i.e., a lipid) in a little capsule, instead of in a dry tablet (i.e., a nonlipid). Water is a **polar** molecule, meaning it has an uneven distribution of electrical charges and is therefore attracted to ions and other polar molecules (see "Why Are Some Molecules Polar?" in Chapter 2 of the text). Lipids do not dissolve in water because they have few functional groups that are polar.

One class of lipids, the fatty acids, is either saturated or unsaturated. Saturated fats are solid at room temperature and tend to raise the level of blood cholesterol in humans. Cholesterol is a lipid necessary for the synthesis of bile acids and steroid hormones. It is also an important component of cell membranes, but too much cholesterol may damage the circulatory system. See Box 3-2 in the text for information about the biochemistry of cholesterol.

Your body can synthesize most of its own fats, including cholesterol, from proteins and carbohydrates. The few it cannot synthesize are easily obtained from whole foods, such as fruits, vegetables, grains, meats, and milk products.

Test for Lipids

1. Take two squares of brown paper. On one, write "vegetable oil;" on the other, write "H_2O."

2. Put one drop of deionized water on the appropriate square. On the other square, place one drop of vegetable oil. Spread the drop with your finger to shorten the drying time. Allow the squares to dry; this will take 20–30 minutes. You can continue with the lab while you wait for the squares to dry.

3. When you hold the dry squares up to a light, the paper will be translucent (light will come through it) where the oil was rubbed in. The oil fills in the spaces between the paper fibers and does not evaporate. This is the positive reaction for this test.

4. Fill out the data sheet at the end of the lab for this reaction. Save the test squares to compare with experimental samples you do later in this lab.

5. If you use a solid substance for this test, such as a nut, take the whole nut and rub it into the paper.

   Q1. Which square is the positive control?

   Q2. Which square is the negative control?

   Q3. Is butter a saturated or unsaturated fat? What about olive oil?

Carbohydrates

Carbohydrates are sugars and the polymers formed by sugars, the polysaccharides. Sugars and polysaccharides are important energy-storage molecules in cells. Many of the carbon atoms in the carbohydrates are attached to hydroxyl groups. Each hydroxyl group can form a hydrogen bond with a water molecule, so sugars and other carbohydrates dissolve in water. See Figure 2-2 for examples of carbohydrates.

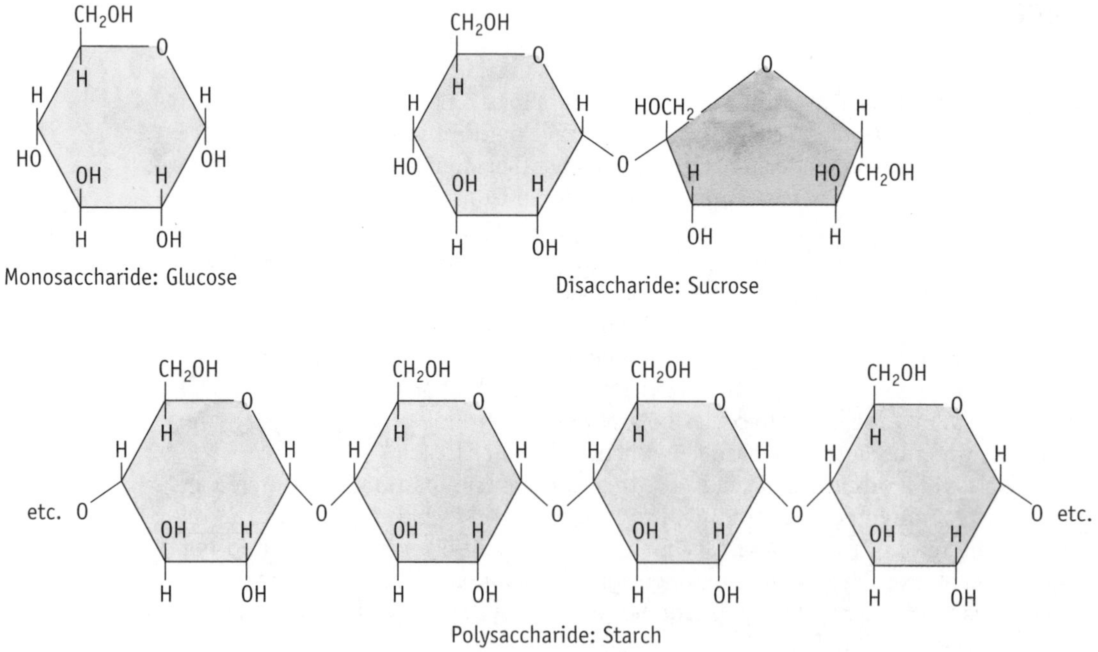

Monosaccharide: Glucose

Disaccharide: Sucrose

Polysaccharide: Starch

Figure 2-2
Examples of Monosaccharide, Disaccharide, and Polysaccharide. Note the numerous hydroxyl (OH) groups that allow hydrogen bonding with water.

The simplest form of sugars are the **monosaccharides** (*mono* = one, *sacchar* = sugar). The most commonly used monosaccharide in biological systems is glucose—this is what plants produce via photosynthesis.

Disaccharides (*di* = two) consist of two monosaccharides joined together. Sucrose, or table sugar, is a disaccharide made of glucose and fructose.

Most plants store sugar in the form of starch, a **polysaccharide** (*poly* = many). Starch is made of many glucose molecules linked together. Another polysaccharide, cellulose, is the main component of plant cell walls and is indigestible by most animals. Dietary fiber is primarily cellulose, as are wood, paper, and cotton.

Almost all carbohydrates in the human diet come from plants. Animals store glucose in the form of glycogen, another polysaccharide.

Test for Monosaccharides and Disaccharides

Materials/Equipment

Hot plate

400 mL beaker

Boiling chips (two or three)

Test tube holder for picking up hot test tubes

Four test tubes

Test tube rack

Deionized water

Benedict's reagent

5% glucose solution

5% fructose solution

5% sucrose solution

Wax pencil and label tape

Safety goggles

Benedict's reagent is the indicator for this test. All monosaccharides have a functional group that includes a double-bonded oxygen atom (see Figure 2-3). When heated, Benedict's reagent reacts with that oxygen, and the color of the reagent changes from blue to green to yellow to red, depending on the amount of sugar present.

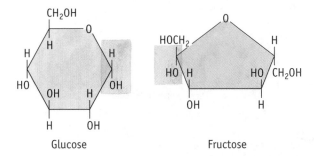

Figure 2-3
Glucose and Fructose. Functional groups that react with Benedict's reagent are highlighted in blue.

Figure 2-4
Boiling Water Bath.

Boiling chips

Figure 2-6
Removing Test Tubes. Handle hot test tubes with a test tube holder.

Wear safety goggles while heating the test tubes in this test.

1. Place two or three boiling chips in the beaker and fill the beaker about halfway with tap water. Put the beaker on the center of the hot plate and turn the hot plate to high, as shown in Figure 2-4. When the water starts to boil, turn down the heat so the water stays at a slow simmer. This is your boiling water bath.

2. Label four test tubes "glucose," "fructose," "sucrose," and "H_2O" using a wax pencil, label tape, or both. Make sure to write high on the test tubes so that the labels do not wash off in the water bath. Into each tube, place 1.5 mL of the appropriate solution: glucose, fructose, sucrose, and deionized water.

3. Add 1.5 mL of Benedict's reagent to each of the samples in the test tubes. Carefully agitate the test tubes so that the contents are well mixed. To do this without splashing, hold the test tube stably in one hand and use the other hand to flick the bottom of the test tube, as shown in Figure 2-5.

4. Note the color of each sample on the data sheet.

5. **Put on your safety goggles.** Point the test tubes away from your face as you place all the test tubes in the boiling water bath. Leave them there for five minutes.

6. **With your goggles on,** remove the test tubes from the water bath using the test tube holder, as shown in Figure 2-6. Do not agitate the tubes. Put them into the test tube rack, and record whether there was a color change on the data sheet.

Q4. Which is the negative control?

Q5. Which is the positive control?

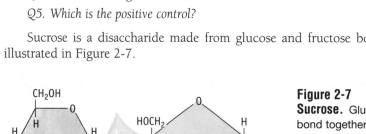

Figure 2-5
Agitating Test Tubes. Method for mixing the contents of test tubes.

Sucrose is a disaccharide made from glucose and fructose bonded together, as illustrated in Figure 2-7.

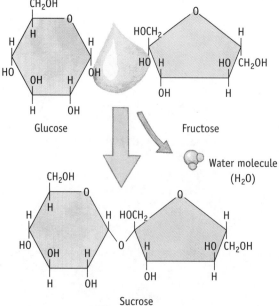

Glucose

Fructose

Water molecule (H_2O)

Sucrose

Figure 2-7
Sucrose. Glucose and fructose bond together and lose a water molecule to form sucrose, a disaccharide.

Q6. Can you explain the results of the test on this sugar, based on its structure? (Hint: Look at the reactive groups on glucose and fructose in Figure 2-3.)

Q7. Would this test work on the disaccharide maltose? Why or why not? Maltose is made from two molecules of glucose bonded together, as shown in Figure 2-8.

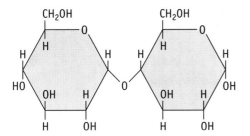

**Figure 2-8
Maltose.**

Test for Starch

Materials/Equipment

Two test tubes

Test tube rack

Lugol's iodine

Starch solution

Deionized water

The indicator for starch is Lugol's iodine. Iodine changes from brownish yellow to bluish black in the presence of starch; in the absence of starch, the indicator remains brownish yellow.

1. Label your test tubes. Put 1.5 mL of starch solution in one and 1.5 mL of deionized water in the other.

2. Add two or three drops of Lugol's iodine to each tube and agitate the tubes gently, as you did for the mono- and disaccharides.

3. Record your results on the data sheet, and save the test tubes for comparison.

Q8. Why does the Benedict's test not work to test for starch?

Proteins

Only 20 amino acid building blocks make up the different types of proteins in all living organisms. The human body contains 50,000–100,000 different proteins. They provide structure, regulate gene expression, control the transport of substances across cell membranes, and function as enzymes for all the chemical reactions in our bodies. A protein molecule's three-dimensional structure determines its function (see Figure 3-18 in the text). Certain chemicals or conditions (such as high temperature) may **denature** a protein by breaking its hydrogen bonds and other weak (noncovalent) bonds. Breaking these bonds alters the protein's three-dimensional shape, causing the properties of the protein to change.

Antibodies (substances that fight infection), hormones (substances that control many developmental and regulatory functions), collagen (a component of skin, muscle, bones, tendons, and ligaments), and keratin (the primary component of hair, horns, nails, and claws) are all proteins.

Eight of the 20 amino acids are the "essential" amino acids. They cannot be synthesized by the human body and must be obtained from foods. See Table 35-1 in the text for a list of the essential amino acids for humans.

Test for Protein

Materials/Equipment

Test tube rack

Two test tubes

0.5% $CuSO_4$

10% NaOH

Egg albumen solution

Deionized water

Gloves

Safety goggles

The indicator in this test is $CuSO_4$, copper sulfate. First, NaOH is added to the sample, which denatures any protein present. In the presence of the denatured protein, the copper in $CuSO_4$ changes from blue to lavender.

NaOH is used in this test to denature the protein: it can also denature the proteins in your skin. Handle NaOH with care, wearing gloves and safety goggles as instructed.

1. Label your tubes "egg albumen" and "H_2O." Egg albumen is egg white, which is mostly protein. The egg albumen solution is egg white mixed with water. Put 1.5 mL of egg albumen into the appropriate tube and 1.5 mL of deionized water into the other.

2. Slowly and carefully add 20 drops of 10% NaOH to each tube. Agitate the tubes gently, using the same method used for the carbohydrates.

3. Add four drops of 0.5% $CuSO_4$ to each tube. Agitate the tubes and place them in the test tube rack. In a few minutes, you should see a color change in the positive control. Record the results on the data sheet and save these tubes for comparison.

Unknown Samples

Now you will use the food samples that you brought or that have been provided. Please share your samples with your classmates. Many food samples have more than one of the organic substances for which you can test. Choose four different samples on which you will perform all the tests. Before you begin, predict which organic substance(s) you think you will find in your samples, and record your hypotheses on the data sheet.

You will need four squares of brown paper and 12 test tubes (three for each of your four samples). Label the paper and test tubes with the sample number and the type of test you will be performing.

- If your sample is a **liquid,** place 1.5 mL of the sample in a test tube.
- If your sample is a **solid,** take a pea-sized portion and chop or break it into the smallest possible pieces using a razor blade or glass stir rod. Put them into a test tube and add 1.5 mL of deionized water.
- If your sample is a **powder** (flour, cornmeal, etc.), put a pinch into 1.5 mL of deionized water in a test tube. Do not add more powder than will stay suspended in the water.

Once **all** of your samples are prepared, return to the instructions for each test and determine which organic substances are in your samples. Divide responsibilities between the members of your group: one person could take a portion of each sample and perform the test for mono- and disaccharides, another person could perform the test for starch, etc. You may compare the controls you did previously for each test with your results, but make sure the colors of the controls haven't changed.

Fill out the table on the data sheet with the results of your tests for each of the four samples.

In this experiment, you made several hypotheses and tested those hypotheses.

Q9. *Which experimental results agreed with your hypotheses?*

Q10. *Which experimental results disagreed with your hypotheses?*

Q11. *Why chop a solid sample into smaller pieces? What reactions would you get from your tests if you did not chop up the solids?*

At the end of the lab

- Pour the contents of your test tubes into the waste containers provided
- Remove any labels from the tubes
- Put the tubes into the designated bin
- Wash your hands

Lab 3   Introduction to Cells
This lab accompanies Chapter 4 of *Asking About Life.*

Materials/Equipment

Compound microscopes

Clean glass slides

Coverslips

Lens paper

Immersion oil

Clear 6-inch rulers

Deionized water with pipettes for making wet mounts

Sterile toothpicks

1% methylene blue, three or four dropper bottles

Sharps disposal

Container for used slides

Autoclave bag

Prepared slides:

    Letter "e"

    Rod-shaped bacteria

Live material:

    Paramecium

    Saccharomyces cerevisiae

    Elodea

    Flower petals (yellow, orange, purple)

Variety of boxes or blocks

Metric rulers

Objectives

1. Learn how to use a compound microscope

2. Understand microscopic scale

3. Use the compound microscope to view cells from different kingdoms, and recognize the different characteristics of those kingdoms

4. Apply the definitions of life to cells

5. Understand the concept of surface area-to-volume ratio

Introduction

Some of the first simple microscopes were made more than 300 years ago by Antonie van Leeuwenhoek, a Dutch cloth merchant (see text). He surprised himself, his friends, and even royalty with his discovery of cells in blood, small creatures swimming in pond water, and other microscopic organisms. Unfortunately, most scientists at the time saw van Leeuwenhoek's findings as mere curiosities. Some 200 years later, technology allowed the manufacture of the good-quality microscopes now used to study cells.

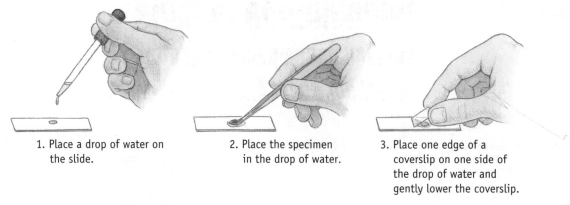

1. Place a drop of water on the slide.

2. Place the specimen in the drop of water.

3. Place one edge of a coverslip on one side of the drop of water and gently lower the coverslip.

Figure 3-1
How to Make a Wet Mount.

You will look at cells from various organisms from several kingdoms. Some microscope slides will be **prepared slides**—that is, the specimen has been preserved so that it will not decay, stained with dyes so that you can see internal structures better, and assembled on a slide with a coverslip. These slides should be returned when you are finished.

For live specimens, make a **wet mount,** as shown in Figure 3-1.

1. Put a drop of water on a slide.

2. Place the specimen in the drop of water.

3. Place one edge of a coverslip on one side of the water drop and gently lower the coverslip, trying not to form air bubbles under the coverslip.

4. When you are finished with the wet mount, dispose of the coverslip and slide as instructed by your teacher.

Compound Microscope

You will be using a **compound microscope,** shown in Figures 3-2 and 3-3, to look at cells. Box 4-1 in the text shows the other types of microscopes used by scientists.

A compound microscope has two sets of lenses that magnify your specimen: the **ocular** (or **eyepiece**) and the **objective lens.** Eyepiece magnification is usually 10X (ten times), depending on the model of microscope. The magnification is written on the tube of the lens.

Q1. What is the magnification of your eyepiece lens?

The objective lens magnifies your specimen further. Your microscope should have three or four objective lenses, which you can rotate into place. **Total magnifica-**

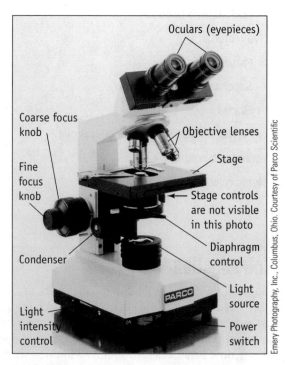

Figure 3-2
Compound Microscope.

Emery Photography, Inc., Columbus, Ohio. Courtesy of Parco Scientific

tion of your specimen equals the magnification of the eyepiece multiplied by the magnification of the objective lens in place.

$$(\text{eyepiece mag.}) \times (\text{obj. mag.}) = \text{total mag.}$$

> *Q2. Determine the possible magnifications of your microscope. Fill in the table on the data sheet.*

The magnifying powers of a microscope alone do not allow you to see objects smaller than those you can see with just your eyes. A microscope must also have good **resolution.** Resolution is the minimum distance between two objects that allows them to be seen as two objects. For instance, a bright spot in the sky may be resolved into two stars with a good telescope; a telescope with less resolution will just magnify the single bright spot. The compound microscope you are using has better resolution than your eyes or a magnifying glass. For even better resolution, an electron microscope can be used (see Box 4-1 in text). Several electron microscope images (micrographs) are used in this lab.

Figure 3-3
Holding a Microscope.
Hold a compound microscope with both hands, supporting its arm and its base.

Using the Microscope

1. Turn on the **light** and adjust the light intensity to a medium setting.
2. Rotate the **coarse focus knob** until the **stage** is as low as possible.
3. Click the **low power objective lens** (scanning power) into place.
4. Clip a prepared slide of the letter "e" onto the stage. Move the slide, using the **stage controls,** until the light shining through the hole in the stage shines through the "e."
5. Look through the eyepieces. If your microscope is binocular, adjust the distance between the eyepieces by gently pushing them together or pulling them apart as you look through them. Experiment until you can see one circle of light with both eyes.
6. **Slowly** move the coarse focus knob until the letter "e" comes into view. Move the slide until the "e" is in the center of your field of view (the circle of light you can see through the eyepieces).
7. Use the **fine focus knob** to bring the "e" into sharp focus. If you have a binocular microscope that has a focusing ring on one of the eyepieces, close your eye on the side with the focusing ring and adjust the fine focus until the image is in focus with your open eye. Close your other eye and use the focusing ring to focus the image for your open eye. Then, open both eyes and look through the microscope. You have adjusted the focus for both eyes.
8. Look at the "e" on the slide itself and then through the microscope. Is it oriented in the same way? Try moving the slide from right to left and up and down, using the stage controls.

> *Q3. How does the image seen through the microscope move relative to how you are moving the actual slide?*

 Do not use the coarse focus when the higher objectives are in place. The space between the slide and the objective lens becomes very small, and you risk smashing them into each other. If your microscope is properly adjusted and your slide is not unusually thick, you will be safe using the fine focus at higher magnifications.

9. Center the "e" in your field of view again. Without changing any settings, rotate the next higher objective lens until it clicks into place.

Q4. What happened to the size of the "e"?

Q5. What happened to the size of the field of view?

Q6. What happened to the intensity of light?

Q7. What happened to the distance between the objective and the specimen?

10. Use the fine focus to focus the "e" and try to adjust the **diaphragm** to change the amount of light coming through your specimen. With specimens on a higher magnification, closing the diaphragm often increases the contrast, allowing you to see internal structures better. Also try adjusting the light intensity.

11. **If the next higher objective is not oil immersion,** perform the following:

    a. Again, without changing any settings, rotate the next higher objective into place.

    Q8. What happened to the size of the "e"?

    Q9. What happened to the size of the field of view?

    Q10. What happened to the light intensity?

    Q11. What happened to the distance between the objective and the slide?

12. **For microscopes with an oil immersion objective,** perform the following:

    You have probably noticed that the light intensity decreases when magnification increases. The oil immersion lens gives a very high magnification, but even more light is lost. Viewing the specimen through oil helps the microscope capture more light.

    a. Go through the steps of getting your specimen into focus, always beginning with the lowest power.

    b. Once the specimen is in sharp focus on the highest "dry" objective, rotate the objective lenses until the specimen is between the high dry and the oil immersion lenses, as illustrated in Figure 3-4.

    c. Put a drop of immersion oil on the coverslip right where the light shines through.

    d. Slowly rotate the oil immersion lens into the drop of oil until it clicks into place.

    e. Carefully adjust the fine focus. It is easy to lose sight of your specimen at this high power.

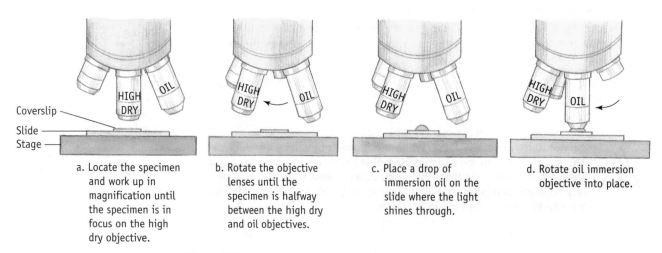

a. Locate the specimen and work up in magnification until the specimen is in focus on the high dry objective.

b. Rotate the objective lenses until the specimen is halfway between the high dry and oil objectives.

c. Place a drop of immersion oil on the slide where the light shines through.

d. Rotate oil immersion objective into place.

Figure 3-4
How to Use an Oil Immersion Objective.

Notes about Oil Immersion

When you use oil immersion, you cannot go back to the dry objective lenses. These lenses can be ruined if oil gets on them. If you decide to use oil, make sure you have finished looking at your specimen at lower powers of magnification.

When you are finished using oil immersion with your slide, lower the stage and remove the slide. If it is a prepared slide, wipe the oil off the slide and the oil immersion lens with **lens paper only.** If it is a wet mount, dispose of the coverslip and slide as instructed by your instructor.

Size of Specimens under a Microscope

Because a ruler does not fit on the microscope with a slide, it cannot be used to directly measure specimens on a microscope. This exercise indicates the size of the field of view so that you have a better idea of the size of your specimens.

1. Return your microscope to the scanning objective and lower the stage. Place a clear metric ruler on the stage with the millimeter (mm) markings over the hole in the stage. Use the coarse focus knob to focus on these markings, placing the millimeter markings so that they go across the diameter of the field of view.

    Q12. How many millimeters across is the field of view?

2. Now, try the next higher objective.

    Q13. How many millimeters across is this field of view?

3. If it is not oil immersion, try the next higher objective. At this magnification, the field of view will be less than 1 mm, so it cannot be measured directly with the ruler. Instead, use the following formula to determine the diameter of this field of view and record the value on the data sheet:

    $$\text{High power diameter} = \text{Low power diameter} \times \frac{\text{Low power magnification}}{\text{High power magnification}}$$

4. Carefully remove the ruler from the microscope and look at the millimeter markings. Scientists use metric units of measurement because they are based on powers of 10. There are 1,000 mm in a meter; in other words, a millimeter is 1/1000 of a meter. Because many specimens are much smaller than a millimeter, they are measured in a metric unit that is 1/1000 of a millimeter (a millionth of a meter!). This unit is called a micrometer and is abbreviated μm. See the Appendix for further explanation of the metric system.

    If your field of view is 6 mm, and there are 1,000 μm in 1 mm, your field of view is 6,000 μm:

    $$6 \text{ mm} \times \frac{1,000 \text{ μm}}{1 \text{ mm}} = 6,000 \text{ μm}$$

5. Convert your measurements of the fields of view into micrometers. Record the values on the data sheet.

    $$\text{Low power: } \underline{\hspace{1cm}} \text{ mm} \times \frac{1,000 \text{ μm}}{1 \text{ mm}} = \underline{\hspace{1cm}} \text{ μm}$$

    Medium power:

    High power (if not oil):

Sizes of some typical cells are shown in Figure 3-5.

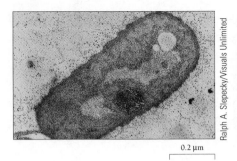

Ralph A. Slepecky/Visuals Unlimited

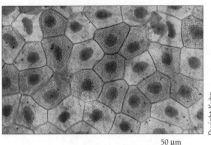

Dwight Kuhn

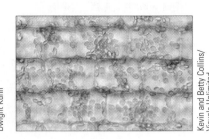

Kevin and Betty Collins/
Visuals Unlimited

|——| 0.2 μm

|——| 50 μm

A. Typical prokaryotic cell: *Bacillus megaterium*, a rod-shaped bacterium.

B. Typical animal cells: Epithelial cells from a salamander.

C. Typical plant cells: *Elodea*, a common aquarium plant.

**Figure 3-5
Sizes of Cells.**

Looking at Cells

Every cell has the following components:

Plasma membrane defines the limits of the cell and regulates the internal environment of the cell (see Figure 4-18 in the text).

Genetic material, in the form of DNA, contains information needed for cell maintenance and for cell reproduction.

Cytoplasm uses information in the DNA to perform cell processes.

When using a compound microscope, always begin on the low power, using the coarse focus knob and fine focus knobs to locate the specimen.

Eukaryotes

The DNA in eukaryotic cells is enclosed in a membrane-bound nucleus (see Figure 3-6). In addition, the cytoplasm of eukaryotic cells includes several membrane-bound organelles for energy production, food storage, protein synthesis, etc. Some eukaryotic organisms are single-celled, and some are multicellular. In multicellular organisms, cells are specialized to perform different functions.

Fill in the blanks in Figure 3-6 (B) with the functions of each structure indicated. See Table 4-1 and Figure 4-5 in the text. You will not be able to see all these structures in your specimens because the resolution is not as high as it is in this image, taken with an electron microscope.

Paramecium

Classified in the kingdom **Protista,** *Paramecium* is a single-celled inhabitant of freshwater ponds.

1. To make a wet mount of *Paramecium,* put a drop of the culture onto a clean slide and place a coverslip on the drop, as shown in Figure 3-1.

2. *Paramecium* moves rapidly using thousands of short, hair-like structures called **cilia,** which cover the surface of the cell and move in an organized fashion. *Paramecium* eats bacteria and small organic particles. Each *Paramecium* cell contains numerous organelles, including vacuoles for food storage and vacuoles for taking up or excreting water. If the organisms are moving too quickly to observe, try removing the coverslip, adding a drop of methyl cellulose (or an appropriate substitute) to the slide, and replacing the coverslip. Methyl cellulose is a thick substance that slows the swimming of the *Paramecium.* It would be like a human trying to swim through honey. Try adjusting the diaphragm to better see the internal structure of the *Paramecium.*

3. Draw two or three cells on the data sheet, labeling all the structures you can.

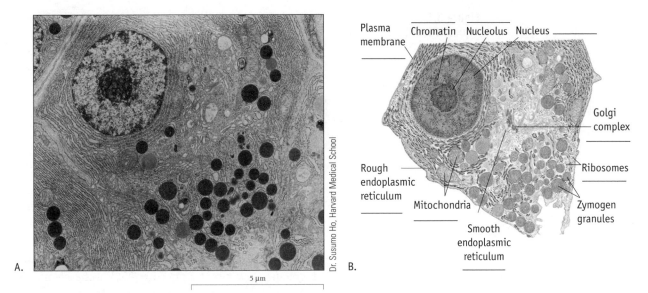

Plasma membrane
Chromatin
Nucleolus
Nucleus
Golgi complex
Rough endoplasmic reticulum
Mitochondria
Ribosomes
Zymogen granules
Smooth endoplasmic reticulum

Dr. Susumo Ho, Harvard Medical School

A.

B.

5 µm

Figure 3-6
Eukaryotic Cell. Electron micrograph of a human pancreas cell (A) and a drawing (B) of the micrograph, showing typical structures in the eukaryotic cell. The zymogen granules contain special digestive enzymes and are not found in other eukaryotic cells.

Saccharomyces Cerevisiae

Saccharomyces cerevisiae is a unicellular organism in the kingdom **Fungi,** an important group of decomposers that includes mushrooms. *Saccharomyces cerevisiae* is a yeast, the type used to make bread, beer, and wine. Through the process of **fermentation,** yeasts convert sugar into carbon dioxide and alcohol. The carbon dioxide causes bread to rise and gives beer its bubbles; the alcohol makes beer and wine "alcoholic."

1. Put a drop of the yeast culture on a clean slide with a coverslip. These are tiny cells, so at low power, focus on a grainy texture. Try adjusting the diaphragm to increase the contrast. As you get to a higher power, you will notice smaller cells attached to larger cells.

2. Yeasts reproduce asexually by **budding** (see Figure 3-7). A cell makes a copy of its DNA (by **mitosis,** a process that will be covered in Lab 8) and packages it into a bit of its own cytoplasm and plasma membrane. A single cell can bud many times.

3. It is difficult to see specific organelles inside the yeast cells because they are small and not stained.

Figure 3-7
Yeast Cell Budding.

Plant Cells

Plants belong to the kingdom **Plantae.** Plants are a major source of oxygen, food, and fuel (gas, oil, and coal). Plant cells have rigid **cell walls** just outside the plasma membrane that keep the cell shape constant. Cell walls have small channels that go through them called **plasmodesmata.**

Q14. Why do plant cells need these channels to live?

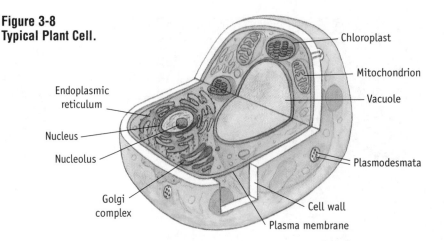

**Figure 3-8
Typical Plant Cell.**

Chloroplast

Mitochondrion

Vacuole

Endoplasmic
reticulum

Plasmodesmata

Nucleus

Nucleolus

Golgi
complex

Cell wall

Plasma membrane

Plant cells also have large **vacuoles,** single membrane-bounded sacs that contain water, wastes, and sometimes pigments. A vacuole can occupy 95% of the volume of a cell, leaving little room for the rest of the cytoplasm (see Figure 3-8).

Elodea
Elodea is a multicellular plant used in freshwater aquariums.

1. Take a leaf of *Elodea* and make a wet mount. When it is in focus under medium or high power, try **through focusing:** move the fine focus knob up and down so you can see through the thickness of the leaf. You are looking at a three-dimensional specimen and should be able to focus on more than one layer of cells.

   Q15. How many cell layers thick is your Elodea *leaf?*

2. Many small, bright-green organelles move around inside each cell. This movement of the cytoplasm and the organelles in it is called **cyclosis.** Cyclosis helps distribute nutrients, gases, and water within the cell. As you through focus, you will notice that these organelles are restricted to the area just inside the plasma membrane. They are not in the center of the cell.

   Q16. What are the green organelles?

   Q17. What structure is taking up the center of the cell?

   Q18. What would you need to do to see the other organelles?

3. Refer to the data sheet for the width of the high-power field of view as you measured it. With the *Elodea* leaf under high power, estimate the length of one cell (in micrometers). For instance, if the diameter was 100 μm, and if there were 10 *Elodea* cells across the field of view, then

$$\frac{100 \ \mu m}{10 \ cells} = 10 \ \mu m/cell$$

   Q19. How many micrometers across is one Elodea *cell?*

4. Draw two or three cells, labeling all the structures you can. Include a scale bar that indicates the length (in micrometers) of one cell.

Flower Petals for Pigments
The bright colors of flowers, fruits, and vegetables come from **pigments.** Some types of pigments are stored in special **plastids;** chlorophyll is stored in chloroplasts. Many orange, yellow, and red pigments are stored in plastids called **chromoplasts.** Some pink and purple pigments are stored not in plastids but in vacuoles. Pigment-containing vacuoles are like bags full of colored water. Have you ever cut a beet? It is different from cutting a carrot. The beet's purple color gets all over the knife, cutting board, and you, if you are not careful. This is because when you cut the beet, you cut open many cells

and their vacuoles, releasing the pigment. On the other hand, the carrot's orange pigment is contained in small plastids, so the color does not pour out of the cells.

1. Tear a small bit of a flower petal and make a wet mount. Focus on the torn edge where the petal is thinnest, hopefully one or two cell layers thick. If the pigment is in the vacuole, the whole cell will appear colored—remember how large the vacuole is? If the pigment is contained in plastids, you will see many small, colored structures within each cell.

   Q20. What type of pigment do you see?

2. Try making a wet mount of a bit of another color of flower petal to see another type of pigment.

3. Draw several cells of each, showing the cell shapes and pigments.

Animal Cells: Human Cheek Cells
You are in the kingdom **Animalia!**

1. Take a clean microscope slide, a coverslip, a sterile toothpick, and a bottle of methylene blue stain.

2. Gently scrape the inside of your cheek with the flat side of the toothpick. Smear the toothpick on the center of the slide and add **one** drop of methylene blue. Put the coverslip on the slide.

3. When you find some cheek cells on low power, center these cells and increase to medium power, then to high power. Notice that like other animal cells, these cells have no cell walls outside the plasma membrane.

   Q21. What cell structure took up the stain?

   Q22. What does this structure contain?

4. Calculate the length of one of your cheek cells.

   Q23. How many micrometers across is one cheek cell?

5. Draw one or two cheek cells.

6. Dispose of this slide in an autoclave bag as soon as you are finished.

Prokaryotes
Prokaryotes do not have an internal membrane system (see Figure 3-9). Instead of being contained in a membrane-bound nucleus, the DNA occupies a region of the cell called the **nucleoid.** Most prokaryotes are microscopic and unicellular. Despite their simplicity of organization, prokaryotes live in almost every environment on Earth.

Bacteria
1. Get a prepared slide of *E. coli* or another bacterium. These are the smallest cells you have examined today. They have been stained with dye so that they are easier to see.

2. Locate the cells at the lowest power, where they will look like colored specks. Focus on a concentrated batch of specks and go to the next higher power. Work your way up to the oil immersion lens, if your microscope has one.

   Q24. You will not see any organelles inside the bacterial cells. Why not?

3. Draw some of the cells you see.

Figure 3-9
Prokaryotic Cell. A prokaryotic cell, such as a cyanobacterium, has a plasma membrane and cytoplasm but no nucleus.

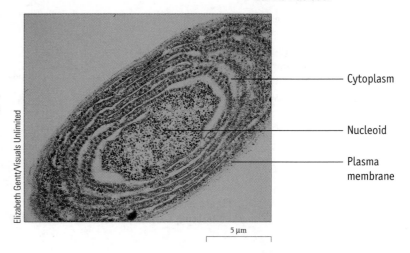

Elizabeth Gentt/Visuals Unlimited

Cytoplasm

Nucleoid

Plasma membrane

5 μm

Surface Area-to-Volume Ratio

All living organisms are composed of cells, and, as you've discovered in this lab, cells are small. The plasma membrane regulates everything that enters or leaves a cell (oxygen, nutrients, waste products, etc.). Although a large cell has more plasma membrane than a small cell, the large cell has proportionally less membrane with which to regulate its volume than the small cell.

The relationship between surface area and volume can be demonstrated by comparing the surface areas and volumes of different-sized boxes or blocks (see Figure 3-10). Record all measurements and calculations on the data sheet at the end of the lab.

1. Choose a small box and a larger box.
2. With a metric ruler, measure the surface area of each box in centimeters.
   a. The surface area of a square or rectangle is the product of its height and its width. A box has six sides. Measure the height and width of each side.
   b. For the total surface area of the box, add together the surface areas of all six sides.
3. Calculate the volume of each box. The volume of a box is a product of its length, width, and height.
4. Now, figure out the surface area-to-volume ratios of your boxes by expressing the two values in a ratio (surface area : volume).

Q25. *Which box had the largest surface area-to-volume ratio?*

Q26. *What does this tell you with respect to cell size? Why are cells so small?*

Q27. *Compare two animals with similar body shapes but different sizes—for example, a cat and a lion. Which would have an easier time maintaining body heat on a cold day? Which would cool down faster in the shade on a cool day?*

Q28. *What kind of leaf type would be the best for a plant that lives in a dry environment (and needs to conserve water from evaporation): leaves with a high or low surface area-to-volume ratio?*

A. Cube

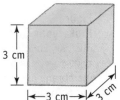

6 sides, each with the same surface area

3 cm

3 cm 3 cm

Surface area = length × width
           3 cm × 3 cm = 9 cm^2
Total surface area = 6 sides × 9 cm^2 = **54 cm^2**
Volume = length × width × height
           3 cm × 3 cm × 3 cm = **27 cm^3**
Surface area-to-volume ratio = 54 : 27
                      = **2 : 1**

B. Rectangular Block

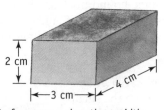

2 cm

3 cm 4 cm

Surface area = length × width

6 sides
$\begin{cases} \text{2 cm × 3 cm = 6 cm}^2 \\ \text{2 cm × 3 cm = 6 cm}^2 \\ \text{2 cm × 4 cm = 8 cm}^2 \\ \text{2 cm × 4 cm = 8 cm}^2 \\ \text{3 cm × 4 cm = 12 cm}^2 \\ \text{3 cm × 4 cm = }\underline{\text{12 cm}^2} \end{cases}$

Total surface area   =   52 cm^2
Volume = length × width × height
        = 4 cm × 3 cm × 2 cm = 24 cm^3
Surface area-to-volume ratio = 52 : 24 = **2.17 : 1**

Figure 3-10
Computing Surface Area-to-Volume Ratios.

Lab 4 — Osmosis and Diffusion
This lab accompanies Chapters 2 and 4 of *Asking About Life.*

Materials/Equipment

Compound microscopes

Clean glass slides

Coverslips

Lens paper

Sharps disposal

Whole milk, 100 mL with disposable pipette

Container for used slides

Objectives

1. Understand and observe Brownian motion

2. Understand and observe diffusion and the factors that affect it

3. Observe and measure the effects of osmosis on cells in solutions of different tonicities

4. Apply the concepts learned to several biological situations

Introduction

One structure common to all cells—from bacterium to plant to animal, from a skin cell to a liver cell to a blood cell—is the **plasma membrane.** See Figure 4-1. The plasma membrane is the boundary between the inside and outside of a cell; it is the site for many biochemical reactions, and it regulates the passage of molecules into and out of cells. For a cell to function properly, it must be able to maintain a fairly constant internal environment, despite changes in the external environment. Changes in the internal environment threaten the three-dimensional structure of the enzymes and other proteins that control all cellular processes. Because the three-dimensional structure of a protein determines its function, such a change could be fatal to the cell.

The plasma membrane is **selectively permeable** because it allows the passage of some molecules but not of others. To understand how membranes regulate the internal environment of cells, it is important to be familiar with the mechanisms by which water and dissolved substances move.

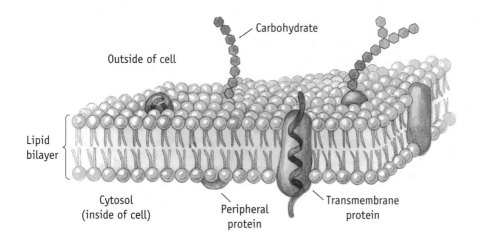

**Figure 4-1
Plasma Membrane.**

Carbohydrate

Outside of cell

Lipid bilayer

Cytosol (inside of cell)

Peripheral protein

Transmembrane protein

Brownian Motion

In 1827, a British botanist named Robert Brown was looking at a wet mount of pollen grains and noticed the pollen grains were jiggling randomly. He thought the pollen grains moved because they were alive, but when he made a wet mount with a dye, the particles of dye (which were clearly not alive) also showed the random motion. A generation later, in 1905, the 26-year-old Albert Einstein explained this phenomenon and named it **Brownian motion.** Brownian motion is the random movement of particles suspended in a fluid (liquid or gas). Liquids and gases are made of tiny molecules in constant motion because of the energy from heat. The warmer the environment, the faster the motion of the molecules. In the case of Brown's pollen grains, the molecules of water were moving and colliding with the pollen grains, causing them to jiggle.

Brownian Motion in Milk

You can easily observe Brownian motion of fat globules in whole milk. Put a drop of milk on a slide with a coverslip and observe it under a compound microscope at 40X.

> Q1. *How would you describe what you see?*

Simple Diffusion

One important result of Brownian motion is diffusion. **Simple diffusion** is the random movement of molecules or ions from an area of high concentration to an area of low concentration. For instance, if you are baking cookies, the cookie smell will start near the oven and diffuse through the air into other rooms. The cookie smell will be strongest in the kitchen, where the concentration of molecules from the baking cookies is the highest. As you move farther from the kitchen, the concentration of molecules decreases, and the smell of cookies diminishes. An area of high concentration next to an area of low concentration like this is called a **concentration gradient.** Concentration gradients are always temporary because diffusion, in addition to the mixing of air, will eventually cause the molecules to become uniformly distributed throughout the system. When the molecules are evenly distributed, the system has reached **equilibrium.** Movement of the molecules will continue, but their overall distribution will not change.

Some substances enter or leave a cell by diffusion across the cell membrane. For instance, when a cell is consuming oxygen (in a process called cellular respiration, which you will study later), there will be a lower concentration of oxygen inside the cell than outside the cell. Oxygen can pass through the cell membrane by diffusion, and as long as there is a concentration gradient, it will continue to enter the cell.

The speed of diffusion depends on several factors, including temperature, the sizes of the molecules, and the extremity of the gradient. Smaller molecules tend to diffuse more quickly; the greater the difference between the two extremes of the gradient, the faster the rate of diffusion.

> Q2. *Do you think a higher temperature causes diffusion to occur more rapidly or more slowly? Explain your hypothesis.*

Temperature versus Rate of Diffusion

To test how temperature affects the rate of diffusion, you will need the following:

Hot plate

Two 400 mL beakers

Two tea bags

Pot holder or oven mitt

1. Fill both beakers halfway with cold tap water.
2. Put one beaker on the hot plate and turn the hot plate to high; leave the other beaker on the countertop.
3. When the water begins to boil, turn off the heat and, using a pot holder or oven mitt, put the beaker of hot water next to the beaker of cold water on your desk. Wait until the water stops moving.
4. Simultaneously lower one tea bag into each of the beakers. **DO NOT MIX.**

   Q3. *Describe how diffusion occurred in the two beakers.*

   Q4. *Which was the experimental test?*

   Q5. *Which was the control test?*

   Q6. *What does this experiment tell you about how temperature affects the rate of diffusion at a molecular level? Was your hypothesis correct?*

Diffusion in a Solid versus Diffusion in a Liquid

As long as there is any amount of heat, the molecules in a substance are in motion—even those in a solid. In fact, diffusion can take place through a solid.

Potassium permanganate is a colored crystal that you can watch diffuse through water and through **agar,** a gelatinous substance derived from marine algae.

   Q7. *Do you think the crystal will diffuse more quickly in the water or in the agar? Explain.*

To test how liquids and solids affect rates of diffusion, you will need the following:

One Petri dish with 2% agar

One Petri dish filled with water

Two crystals of potassium permanganate (handle with forceps)

Metric ruler

One piece of white paper

1. Put the dish with water and the dish with agar on the piece of white paper.
2. When the water stops moving in its dish, put one crystal of potassium permanganate in the center of each dish. Start timing diffusion now.
3. Every 3 minutes for 15 minutes, measure the diameter (in millimeters) of the pinkish purple permanganate in each dish. Record the measurements on the data sheet.
4. Continue to measure and record the diameter of the permanganate in both dishes every 15 minutes (while you continue with the lab) for the rest of the period.
5. Plot both sets of data on the same graph so that you can compare the rates of diffusion that took place in the water and in the agar. Refer to the Appendix if you need help graphing. Draw each curve in a different color or with a different type of line (for instance, solid or dashed). Label each curve.

   Q8. *On your graph, which is the independent variable: time or diameter?*

   Q9. *Did diffusion take place in both dishes?*

   Q10. *Did the potassium permanganate diffuse more quickly in one dish than in the other? Why? Explain what happened on a molecular level.*

Osmosis

Water is a **solvent,** a fluid in which other substances dissolve. The substances that dissolve in a solvent are called **solutes.** Although water and some small molecules and ions can move freely across a selectively permeable membrane, many solutes cannot.

When the concentration of solutes on one side of a selectively permeable membrane is unequal to the concentration of solutes on the other side of the membrane, a concentration gradient exists. If the solutes cannot move across the membrane, water will move instead. The movement of water molecules through a selectively permeable membrane in response to a concentration gradient is a type of diffusion called **osmosis.**

If a red blood cell is placed in a beaker with a solution that has a higher concentration of solutes than the concentration inside the cell, water will follow the concentration gradient and leave the cell. The cell will shrivel as it loses water. This is called **hypertonic solution:** the solution contains a higher concentration of solutes than the cell. (*Hyper* = above. For instance, if you are hyperactive, your activity level is above normal.)

A **hypotonic solution** is a solution that has a lower concentration of solutes than the concentration inside the cell. (*Hypo* = below. For example, a hypodermic needle goes below the dermis, or skin.) If a red blood cell is placed in a beaker of water, the water molecules will follow the concentration gradient and move into the cell. The cell will swell and eventually burst.

A solution with a concentration of solutes equal to the concentration inside the cell is called an **isotonic solution** (*iso* = the same). Blood cells placed in an isotonic solution will retain their shape. See Figure 4-17 in the text for photos of blood cells in different solutions. The terms hypotonic, hypertonic, and isotonic refer to the **tonicity** of a solution (see Figure 4-2).

You will determine the tonicity of several salt solutions (salt dissolved in water) compared with a chunk of eggplant, which is made of many cells. The combined responses of the cells to the solutions will be great enough to weigh.

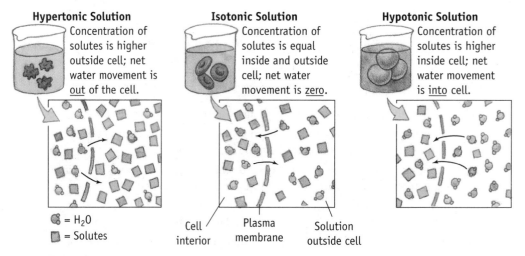

Figure 4-2
Red Blood Cells in Solutions of Different Tonicities.

Materials per Group

Four cubes of eggplant, about 8 g each (3×3×3 cm), without skin

Four 100 mL beakers

About 50 mL each of solutions A, B, C, and X

Balance

Weight boat

Wax pencil

Part A

1. Label three of your beakers "A," "B," and "C." Fill each approximately halfway with the appropriate solution.

2. Weigh three of the eggplant cubes to the closest 100th of a gram.

   a. Obtain a **weight boat** (a small plastic tray). Weigh the weight boat on the balance to the closest 100th of a gram. Record this weight.

   b. Place a cube of eggplant into the weight boat, place the weight boat on the balance, and measure the weight to the closest 100th of a gram. Subtract the weight of the weight boat to determine the weight of the eggplant cube. Record this weight on the data sheet.

   c. Weigh the other two eggplant cubes using the same method. Track the weight of each cube.

3. Noting which eggplant cube is which, place each into one of the labeled beakers—perhaps, write the cube's starting weight on the beaker label.

4. Write on the data sheet the time that you start soaking the eggplant cubes, and allow the cubes to soak about 45 minutes. While you wait, go to Part B.

5. After 45 minutes, gently dry each eggplant cube with a paper towel (do not squeeze the cube, or you will squeeze out water, which will change its weight) and weigh it, using the weight boat. Record the weights on the data sheet.

   Q11. Based on your results, what is the tonicity of each solution?

   Q12. What is happening to cause the weights of the eggplant cubes to change?

   Q13. Describe the appearance and texture of the three cubes of eggplant after soaking them in the solutions.

Part B

1. Label your last beaker "X" and fill it halfway with solution X.

2. Weigh the last eggplant cube to the closest 100th of a gram and place it into solution X. Record the weight and time on the data sheet.

3. Carefully dry and reweigh the cube every 5 minutes for 30 minutes.

4. Plot your results on a graph to visualize the rate of water movement into or out of the cells.

   Q14. Which is the independent variable: weight or time? Which is the dependent variable?

   Q15. What is the tonicity of solution X?

   Q16. Referring to your graph, is the rate of water loss or uptake constant over the 30 minutes?

Lab 5   Enzymes

This lab accompanies Chapter 5 of *Asking About Life.*

Materials/Equipment

Water baths for test tubes

Ice bath with thermometer

37°C water bath

Boiling water bath

Label tape (optional)

Wax pencils

Waste beaker

Dish tub for dirty test tubes

Thermometer for measuring room temperature

Deionized water in three or four flasks with 3 mL pipettes

Razor blades

Sharps disposal

pH solutions, 50 mL flasks each of pH 3, pH 7, and pH 11 with 3 mL pipettes

Small cubes (about 1 cm^3) of apple, potato, chicken, and carrot (about 10 each)

Forceps for handling cubes

Per group of three or four students:

    40 mL 3% H_2O_2 with 3 mL pipette

    20 mL liver puree with 3 mL graduated pipette

    15 standard test tubes

    Test tube holder

    Test tube rack

    Glass stir rod (long enough to reach to the bottom of the test tubes)

    Forceps

Objectives

1. Understand the importance and function of enzymes

2. Predict and test the activity of the enzyme catalase under different temperatures, concentrations, and pH levels and in different types of tissues

3. Apply knowledge gained from the experiments to various biological and food storage situations

Introduction

Even when you are sitting still, the 10 trillion living cells of your body are busy. Muscle cells in your heart are contracting; red blood cells are picking up oxygen from your lungs and distributing it throughout your body. Many cells are growing and dividing. All of the cells are performing the functions necessary to keep themselves alive: breaking down sugars, disposing of wastes, building membranes, producing enzymes, etc. All the chemical reactions that accomplish these activities require energy.

According to the **first law of thermodynamics,** energy cannot be created or destroyed; it can only be changed from one form to another. So, cells must use the energy from **exergonic** (energy releasing) reactions to power the **endergonic** (energy-consuming) reactions. To synchronize these two types of reactions, cells use large molecules, usually proteins, called **enzymes.** Enzymes have several characteristics:

- Enzymes speed reactions between other molecules—without them, most reactions would occur too slowly to sustain life.
- Enzymes speed reactions by lowering the **activation energy** (the minimum energy required for a reaction to occur) of reactions. They accomplish this by participating directly in the reaction, by orienting the substrate molecules to increase the chances of their interaction, or by straining the bonds of the substrate to pave the way for new bonds.
- Enzymes are not changed by the reactions they speed.
- Enzymes are specific: each enzyme lowers the activation energy for one type of reaction.

As with all proteins, the three-dimensional shape of an enzyme molecule is critical to its performance. Part of its structure includes an **active site,** a place where the enzyme binds to reacting molecules called **substrates.** It is this binding that lowers the activation energy (see Figures 5-1 and 5-2). If the shape of the enzyme is altered, the active site will not fit and bind to its substrate.

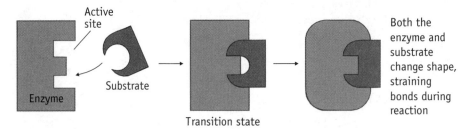

Figure 5-1
Induced Fit Model of Enzyme–Substrate Interaction. The enzyme and the substrate each change shape, straining bonds during the reaction. The strained form of a substrate—the transition state—exists for as little as a billionth of a second.

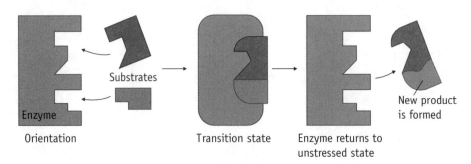

Figure 5-2
How Does an Enzyme Catalyze a Reaction between Two Molecules? The enzyme brings both substrates close together in the proper orientation and strains the covalent bonds of the substrates, lowering the activation energy of the reaction.

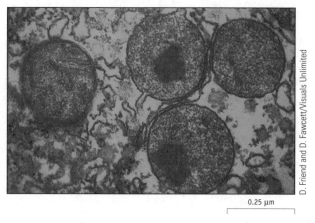

D. Friend and D. Fawcett/Visuals Unlimited

0.25 µm

Figure 5-3
Peroxisome. Peroxisomes contain enzymes that degrade molecules in reactions that produce hydrogen peroxide.

The liver makes many different enzymes. Some are metabolic, converting fructose and other sugars into glucose or making glucose and proteins into glycogen for storage. Some of the enzymes convert toxic compounds such as alcohol and nitrates into less toxic compounds or into compounds that can be easily cleared from the body. Hydrogen peroxide (H_2O_2) is a product of many of these reactions. It is highly reactive and can easily damage or kill cells if left intact. Cells use the enzyme **catalase** to break the hydrogen peroxide into harmless products. In a cell, catalase is contained in a special type of organelle called a **peroxisome**, shown in Figure 5-3.

The reaction you will be testing is the conversion of hydrogen peroxide into water and oxygen gas by the enzyme catalase:

$$2\ H_2O_2 \xrightarrow{\text{catalase}} 2\ H_2O + O_2$$

Q1. What is the substrate in this reaction?

Because hydrogen peroxide is a byproduct of many biochemical reactions, many cells have catalase activity. The liver, with its detoxifying functions, has higher levels of catalase than most cells. When hydrogen peroxide is added to ground beef liver, there will be vigorous bubbling.

Q2. Production of what gas causes the bubbling?

 For all these experiments, use a number scale from 0 to 5 to describe the level of activity, with 0 = no reaction and 5 = vigorous bubbling.

Control

1. Take two test tubes: label one "positive control" and the other "negative control."

2. Measure 1 mL of liver puree with the pipette: insert the pipette into one of the test tubes, and lower the tip as close as possible to the bottom of the test tube before squeezing out the liver puree. If liver is smeared on the sides of the tube, the amount available for reaction will not be consistent among your tests.

3. Put 1 mL of liver puree into the other tube, using the same method.

4. Into one test tube (which one?), put 3 mL of deionized water. Describe the results on the data sheet.

5. Into the other test tube, put 3 mL of H_2O_2. Watch the reaction for several seconds, noting the height the bubbles travel as well as the speed at which the reaction occurs. Assign this level of bubbling "4." **These will be the standards to which you compare the other reactions.**

6. According to the **second law of thermodynamics,** the amount of energy available to do work decreases in the course of any process. Often, the energy unavailable to do work is in the form of heat.

   Q3. Feel the test tube with your hand. Did this reaction give off heat?

7. Describe your results on the data sheet.

Effect of Temperature on Enzyme Activity

Increasing temperature generally increases the effectiveness of enzymes because, as the enzyme and substrate molecules move more rapidly, their chances of colliding and interacting increase. However, as with all proteins, the three-dimensional structure is in danger when the temperature gets high enough to break weak bonds. When bonds are broken and the enzyme loses its three-dimensional structure, the enzyme is **denatured:** it loses its activity.

Every enzyme has a range of temperatures at which it is the most effective. Most enzymes in the human body work best at body temperature, about 37°C.

1. Using the thermometer at the front of the classroom, find the **room temperature** of the classroom in degrees Celsius. Record the number: _____°C.

2. Check the thermometer in the **ice bath** and record its temperature: _____°C.

3. Put 1 mL of liver puree into four clean test tubes using the method described in Step 2 of the previous control experiment. Label each tube with your group's name and the following temperatures:

   Tube 1: temperature of the ice bath
   Tube 2: room temperature of the classroom
   Tube 3: "37°C"
   Tube 4: "100°C"

4. Into four more clean test tubes, put 3 mL of H_2O_2. Label each tube with your group's name and the same temperatures as the liver tubes.

5. Leave the room temperature tubes of liver and H_2O_2 in the test tube rack on your desk. Put each of the other six tubes into the appropriate water baths. After four minutes, **use the test tube holder to remove the tubes from the boiling water bath** and bring them back to your desk.

6. Add each tube of H_2O_2 to the corresponding tube of liver. Rate the reactions and record your results on the data sheet.

7. Make a graph of the activity rate vs. the temperature on the data sheet.

   Q4. Is there a temperature at which the catalase activity was the most effective? Was it the temperature you were expecting? If it was not, how can you explain the results?

   Q5. Why put the tubes of H_2O_2 (not just the liver tubes) into the water baths?

Effect of pH on Enzyme Activity

The pH of a solution is a measure of how acidic or basic it is. Pure water with a neutral pH (in which the H^+ and OH^- concentrations are equal) has a pH value of 7. An acidic solution has a higher concentration of H^+ ions and a lower pH value (<7); a basic solution has a lower concentration of H⁺ ions and a higher pH (>7). See Chapter 2 of the text for more details.

Enzymes usually have a narrow range of pH values at which they work best, just like they have an optimum temperature range. The pH may alter the shape of the enzyme, or it may add or remove hydrogen ions from the enzyme, changing the way it binds to the substrate.

Q6. In which pH range do you suppose catalase works best?

Q7. On the graph on the data sheet, draw a curve that corresponds to your prediction.

1. Label three test tubes "pH 3," "pH 7," and "pH 11."
2. Carefully put 1 mL of liver puree into each tube.
3. Add 2 mL of the appropriate pH solution to each test tube.

 Acids and bases can burn your skin and damage your clothing. Handle these solutions with care. Wear gloves and eye protection as instructed.

4. Wait about 20 minutes. While you are waiting, continue with the next activity.
5. Add 3 mL of H₂O₂ to each tube. Rate the reactions and record your results on the data sheet.

Q8. At what pH does catalase exhibit the greatest activity?

Q9. Draw a new curve on your graph that corresponds to the results of your test.

Effect of Enzyme Concentration on Enzyme Activity

In this experiment, you will rate catalase activity at full strength (100%), 10%, and 1%.

Q10. What results do you expect in this experiment?

Q11. What is the explanation, at the molecular level, for your prediction?

1. Get five clean test tubes. Label the tubes "100%," "10%," "1%," "stock 10%," and "stock 1%."
2. Into the "100%" tube, put 1 mL of liver puree, carefully placing all liver puree at the bottom of the tube.
3. Into the "stock 10%" tube, put 1 mL of liver puree. Add 9 mL of deionized water and mix by squeezing the pipette several times. You have now diluted the liver to 10% of the original concentration. Put 1 mL of this mixture into the "10%" tube.
4. Into the "stock 1%" tube, put 1 mL of the "stock 10%" solution and (specify how much) _____ mL deionized water. Mix with the pipette; put 1 mL of this mixture into the "1%" tube.
5. Add 3 mL of H₂O₂ to each of the following tubes: "100%," "10%," and "1%." Rate the reactions and record results on the data sheet.

Q12. Were the results of this experiment as you expected? If they were not, can you explain the results?

Q13. How effective is catalase? In this experimental setup, do you think you would detect a difference in reaction rates of a 100% and a 50% liver solution?

Catalase Activity in Different Tissues

1. Label five test tubes "liver," "chicken," "apple," "potato," and "carrot."

    Q14. Which of these do you predict will exhibit catalase activity?

2. Put 1 mL of liver puree into the "liver" tube. For each of the other samples, get a piece about 1 cm³, equivalent to the volume of the 1 mL of liver. Use a razor blade to **carefully** cut each sample into smaller pieces. Add the chopped samples to the appropriate tubes. You may push them down using a glass stir rod. Carefully wash the razor blade (or get a new one) and the stir rod between samples.

 Wash your hands thoroughly with soap and water immediately after handling raw chicken.

3. Add 3 mL of H_2O_2 to each tube. Watch for several seconds, rate the reactions, and record your results on the data sheet.

    Q15. What do your results tell you about the functions of the different types of tissues? Describe your results for each tissue tested and your explanations for each result.

    Q16. Why was it important to wash the razor blade and stir rod between samples?

    Q17. Why was it important to cut the samples into smaller pieces?

When you finish this lab

*   Dispose of the contents of your test tubes into the waste beaker
*   Remove the labels from your test tubes
*   Place the empty test tubes into the wash bin
*   Wash your hands

Enzymes
Data Sheet

Introduction

Q1. *What is the substrate in the reaction that converts hydrogen peroxide into water and oxygen gas by the enzyme*

catalase? _____

Q2. *Production of what gas causes the bubbling in ground beef liver?* _____

Control

Q3. *Feel the test tube with your hand. Did this reaction give off heat?* _____

	ENZYME ACTIVITY (1-5)	DESCRIPTION
Positive control		
Negative control		

Effect of Temperature on Enzyme Activity

TEST TUBE	TEMPERATURE (°C)	ENZYME ACTIVITY (1-5)
Tube 1 (ice bath)		
Tube 2 (room temperature)		
Tube 3 (body temperature)		
Tube 4 (boiling water)		

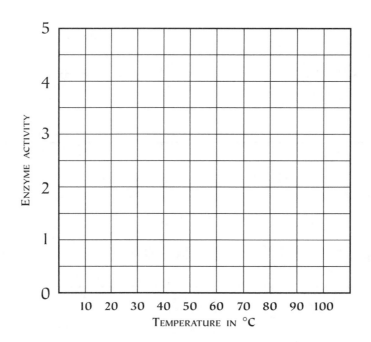

Q4. *Is there a temperature at which the catalase activity was the most effective? Was it the temperature you were expecting? If it was not, how can you explain the results?* _____

_____

_____

Q5. Why put the tubes of H_2O_2 (not just the liver tubes) into the water baths? _____

_____

_____

Effect of pH on Enzyme Activity

Q6. At which pH range do you suppose catalase works best? _____

_____

_____

Q7. On the graph, draw a curve that corresponds to your prediction.

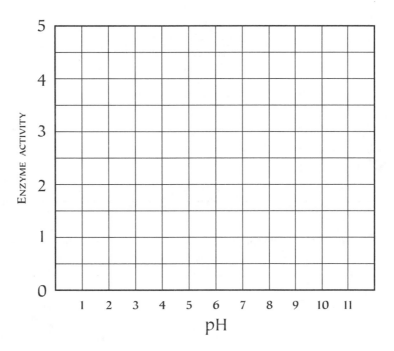

Fill in the table with your results.

	ENZYME ACTIVITY
pH 3	
pH 7	
pH 11	

Q8. At what pH does catalase exhibit the greatest activity? _____

Q9. Draw a new curve on your graph that corresponds with the results of your test.

Effect of Enzyme Concentration on Enzyme Activity

Q10.   What results do you expect in this experiment? _____

_____

_____

Q11.   What is the explanation, at the molecular level, for your prediction? _____

_____

_____

ENZYME CONCENTRATION	ENZYME ACTIVITY	DESCRIPTION
100%		
10%		
1%		

Q12.   Were the results of this experiment as you expected? If they were not, can you explain the results? _____

_____

_____

Q13.   How effective is catalase? In this experimental setup, do you think you would detect a difference in the reaction rates

of a 100% and a 50% liver solution? _____

_____

_____

Catalase Activity in Different Tissues

Q14. *Which of these tissues—liver, chicken, apple, potato, and carrot—do you predict will exhibit catalase activity?*

_____

TISSUE	ENZYME ACTIVITY	DESCRIPTION
Liver		
Chicken		
Apple		
Potato		
Carrot		

Q15. *What do your results tell you about the functions of the different types of tissues? Describe your results for each tissue tested and your explanations for each result.* _____

_____

_____

Q16. *Why was it important to wash the razor blade and stir rod between samples?* _____

_____

_____

Q17. *Why was it important to cut the samples into smaller pieces?* _____

_____

_____

Questions

Q18. *List the conditions you tested under which catalase exhibited the greatest activity. How do these conditions compare with those of a cell in a human body?* _____

_____

_____

Q19. *H_2O_2 is commonly used as a disinfectant for scrapes and cuts. What are you trying to do when you apply H_2O_2 to your scraped knee? What on your knee causes the H_2O_2 to bubble when you apply it?* _____

_____

_____

The browning that occurs when fresh potatoes and apples are cut is a result of the following reaction:

$$\text{Catechol} + O_2 \xrightarrow{\text{catecholase}} \text{Benzoquinone}$$
$$\text{(colorless)} \qquad\qquad\qquad \text{(reddish brown)}$$

Q20. *Why do mashed potatoes stay white?* _____

_____

_____

Q21. *Some people put fresh lemon juice on fruit salad to keep it from browning. What might be the chemical explanation for this practice?* _____

_____

_____

The U.S. Food and Drug Administration recommends that uncooked beef be refrigerated no more than 3–5 days before it's eaten; for uncooked fish, they recommend only 1–2 days.

Q22. *Why do you think fish might not keep as long in the refrigerator?* _____

_____

_____

It also is recommended that uncooked ground beef be refrigerated only 1–2 days. Why might this be? _____

_____

_____

Q23. *Fever is a common symptom of a viral or bacterial infection. What are two functions of a rise in body temperature in this case? What would the danger be if the temperature got too high (above 108°F, or 42°C, in humans)?* _____

_____

_____

Lizards and snakes often sit in sunny spots (on exposed rocks, in the middle of a road) in the morning. They do not use the heat generated by their bodies to heat themselves; they obtain heat from the environment. After a cold night, they are sluggish and must heat up before they can be active.

Q24. *Why do they need heat?* _____

_____

_____

Q25. *Where does the body heat of mammals and birds come from? (The second law of thermodynamics might help you answer this one.)* _____

_____

_____

Lab 6   Cellular Respiration

This lab accompanies Chapter 6 of *Asking About Life.*

Objectives

1. Understand the process that converts food energy into energy cells can use

2. Understand and identify reduction and oxidation reactions

3. Distinguish between aerobic and anaerobic reactions

4. Identify the products of alcoholic fermentation

5. Analyze the results of a test for the citric acid cycle in hamburger mitochondria

Introduction

Plants do not need to eat. They obtain energy from the sun and use this light energy to produce chemical energy in a process called **photosynthesis,** the subject of the next lab. Organisms such as plants that synthesize their own organic molecules from inorganic material are **autotrophs** (*auto* = self, *trophe* = nourish). Like fungi and many other organisms, humans are **heterotrophs** (*hetero* = other). We must obtain our energy from other organisms by eating autotrophs, other heterotrophs, or both.

All cells use ATP (adenosine triphosphate) to fuel their biological processes. Both heterotrophs and autotrophs obtain ATP by breaking down organic molecules in a process called **cellular respiration.**

Cellular respiration begins with small organic molecules, such as the sugar glucose. If you eat a piece of pizza, cellular respiration cannot begin until you have **digested** the pizza, or reduced it to smaller units. Proteins, polysaccharides, and lipids must be broken into smaller molecules that may enter a cell. Cellular respiration may then occur: the energy in the amino acids, monosaccharides, and fatty acids is converted into the chemical bonds of ATP.

During cellular respiration of glucose, the bonds between the carbon atoms in glucose are broken, and the energy stored in those bonds is used to form the high-energy bonds in ATP. In the process, the glucose molecule is broken down into carbon dioxide and water.

$$C_6H_{12}O_6 + 6\ O_2 \xrightarrow{\text{cellular respiration}} 6\ CO_2 + 6\ H_2O$$
$$\text{glucose + oxygen} \qquad\qquad \text{carbon dioxide + water}$$

The complete breakdown of glucose occurs in the four stages shown in Figure 6-1: **glycolysis,** the formation of **acetyl-CoA,** the **citric acid cycle,** and the **electron transport chain.** Glycolysis occurs in the cytosol of the cell; the rest of cellular respiration occurs in the mitochondria.

In this lab, you will perform experiments to identify products formed during two of the four stages of cellular respiration. Many different enzymes work to catalyze the numerous reactions that occur in cellular respiration. Two **coenzymes,** NAD^+ and FAD, also participate in these reactions. Unlike enzymes, coenzymes are not catalysts for chemical reactions; instead, they are carrier molecules. These coenzymes take electrons from glucose and eventually transfer those electrons to oxygen. **Oxidation** is the loss of an electron by a molecule, ion, or atom: a molecule, ion, or atom that loses an

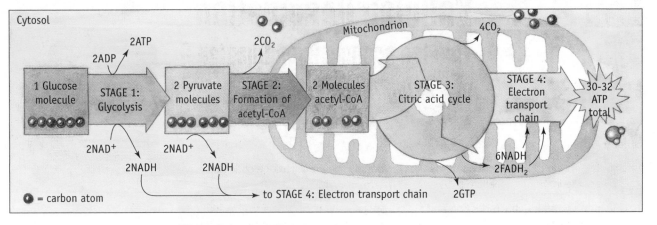

Figure 6-1
Overview of Cellular Respiration.

electron is **oxidized. Reduction** is the gain of an electron by a molecule, ion, or atom: one that accepts an electron is **reduced.** The electron transfer to oxygen in the electron transport chain is coupled with the formation of ATP.

> ⚠️ **Some of these experiments take time.** Set up the carbon dioxide production experiment first, then the citric acid cycle experiment. Do the ethanol production experiment last.

> ⚠️ **Safety:** Because you will be handling yeast cultures and raw meat (hamburger), wash your hands periodically and do not eat or touch your mouth or face while in the lab.

Glycolysis

The first stage in cellular respiration involves the breakdown of glucose, a six-carbon sugar, into two molecules of **pyruvate,** each with three carbon atoms. In this process, shown in Figure 6-2, some energy stored in the glucose is converted into two high-energy phosphate bonds in ATP. In addition, two molecules of the coenzyme NAD^+ are reduced: they each accept two electrons and a hydrogen ion to become two high-energy molecules of NADH.

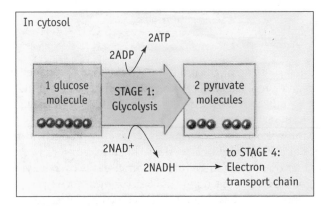

Figure 6-2
Glycolysis. Stage 1 of cellular respiration.

In the presence of oxygen, NADH is oxidized back to NAD$^+$. It transfers its new electrons to the electron transport chain for the formation of ATP. Then, the NAD$^+$ helps break down another molecule of glucose into pyruvate, continuing the process of glycolysis.

Fermentation

In the absence of oxygen, the cell still must oxidize its NADH, or glycolysis cannot continue. Have you ever exercised vigorously and experienced a tingling in your muscles? Your muscle cells needed more oxygen than your blood and lungs could deliver and had to oxidize NADH to continue glycolysis to get energy. In the absence of oxygen, NADH can donate its electrons to pyruvate, converting the pyruvate into **lactic acid** in a process called **lactic acid fermentation.** The accumulation of lactic acid in your muscles caused the tingling. As oxygen becomes available, the lactic acid converts back to pyruvate.

$$\text{pyruvate} \xrightarrow{\text{no oxygen}} \text{lactic acid} \xrightarrow{\text{oxygen}} \text{pyruvate}$$

$$\text{NADH} \longrightarrow \text{NAD}^+$$

NAD$^+$ can now participate in the breakdown of another molecule of glucose, allowing glycolysis to continue.

Although animals, like humans, can make up for a temporary lack of oxygen by forming lactic acid, many microorganisms normally extract energy in the absence of oxygen. Such microorganisms are **anaerobic.** They also undergo fermentation, but the end products depend on the organism. For instance, yeast cells do not produce lactic acid from pyruvate; instead, they form ethanol (alcohol) and carbon dioxide from pyruvate during the oxidation of NADH. This process is called **alcoholic fermentation.**

$$\text{pyruvate} \xrightarrow{\text{no oxygen}} \text{ethanol} + \text{carbon dioxide}$$

In the following two experiments, you will identify the production of carbon dioxide and ethanol by yeast during alcoholic fermentation. This is the same type of yeast you saw budding in Lab 3, which discusses cells. The yeast cultures you will be using have been grown in a solution containing glucose.

Carbon Dioxide Production

In this experiment, you will set up a **respirometer,** a simple device for measuring the amount of carbon dioxide produced during respiration.

Materials/Equipment

24-hour yeast culture

Deionized water in 250 mL flask with pipette

5% glucose in 250 mL flask with pipette

5% sucrose in 250 mL flask with pipette

37°C water bath with test tube rack

Per group of two or three students:
    Three large test tubes
    Three small test tubes
    Label tape and markers
    Metric rulers

1. Practice setting up the respirometer, as shown in Figure 6-3.

    a. Fill a small test tube completely with tap water.

b. Invert a large test tube over the small one. Use your finger or the blunt tip of a pencil to push the small tube up until its rim is in contact with the bottom of the large tube.

c. **Quickly** invert the tubes so as little water as possible leaks from the small tube. A small bubble will form in the small test tube.

d. Repeat this procedure until you can make a respirometer with the smallest bubble possible.

Figure 6-3
Setting Up a Respirometer.

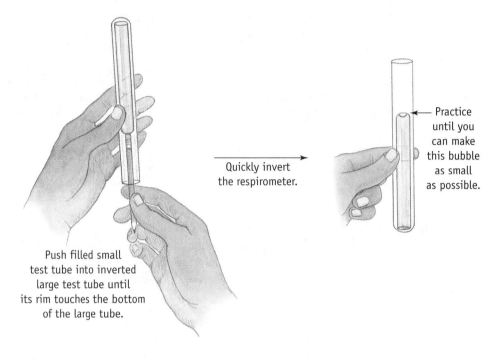

Push filled small test tube into inverted large test tube until its rim touches the bottom of the large tube.

Quickly invert the respirometer.

Practice until you can make this bubble as small as possible.

2. Use small pieces of tape to label your three small test tubes "control," "glucose," and "sucrose." Recall that sucrose is a disaccharide composed of a glucose molecule linked to a fructose molecule.

3. Mark one small test tube at the 2/3 full level. Use this tube to mark the 2/3 level on the other two small tubes. Fill each tube to this level with the yeast suspension, as shown in Figure 6-4.

4. Fill each small tube to the top with the appropriate solution. With what should you fill the control tube?

5. Using the technique you practiced, invert each small tube into the empty large tubes so that you have three respirometers.

6. Mark each large tube at the starting level of the air bubble, as illustrated in Figure 6-5. Place your respirometers into a 37°C water bath. Allow them to ferment at least an hour while you continue the lab. Check them periodically. As the yeast ferments, the gas bubbles at the top of the small tubes will grow.

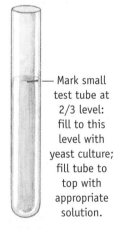

Mark small test tube at 2/3 level: fill to this level with yeast culture; fill tube to top with appropriate solution.

Figure 6-4
Filling the Small Test Tube.

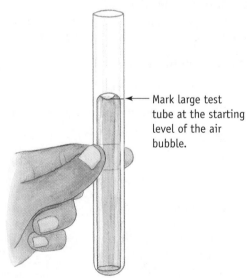

Mark large test tube at the starting level of the air bubble.

Figure 6-5
Marking the Starting Level on the Respirometer.

Q1. What is responsible for the increasing space at the top of the small tubes?

Q2. What is your hypothesis about the outcome of this experiment?

Q3. Explain your hypothesis.

7. After at least an hour, remove the respirometers from the water bath and measure the distance between the final level of the yeast culture and your original mark. Record these measurements on the data sheet.

Q4. How do your results compare with your original hypothesis?

Q5. What difference did you expect between the results in the glucose tube and those in the sucrose tube? What happened? What do you think would happen if you allowed the tubes to ferment longer? Explain your idea.

Q6. Why do you want the original bubble in the respirometer to be as small as possible? What is in this bubble that will interfere with your results?

Ethanol Production

Do this experiment after you set up the next experiment—citric acid cycle.

Ethanol is the other product of fermentation. The indicator used in this experiment is Lugol's iodine. When alcohol (clear) and iodine (brownish yellow) are combined in the presence of sodium hydroxide (NaOH), a cloudy yellow precipitate, **iodoform,** is formed.

 Sodium hydroxide is extremely caustic. Exercise caution and wear gloves and goggles when using this solution.

Materials/Equipment

24-hour yeast culture, settled, with pipette (attached to side of flask)

Deionized water in three or four flasks with pipettes

10% NaOH in three or four small flasks with pipettes

70% ethanol in 50 mL flask with pipette

Gloves and goggles for handling NaOH

Per group of two or three students:
Three test tubes
Lugol's iodine in dropper bottle
Test tube rack

1. Label your tubes "1," "2," and "3."

2. Tube 1

   a. Put 1.5 mL of deionized water into the tube.

   b. Add 1.5 mL of ethanol.

   c. Slowly add 1 mL of NaOH down the side of the tube and mix by tapping the tube with your finger.

   d. Slowly add about 10 drops of Lugol's iodine and look for the formation of a cloudy yellow precipitate, iodoform. Record your results on the data sheet.

3. Tube 2

   a. Carefully pipette 3 mL of the clear solution above the settled yeast cells from the flask into your test tube. Do not mix the yeast culture.

   b. Slowly add 1 mL of deionized water down the side of the tube and mix by tapping the tube with your finger.

   c. Slowly add about 10 drops of Lugol's iodine. Record your results on the data sheet.

4. Tube 3

    a. Carefully pipette out 3 mL of the clear solution above the settled yeast cells into the test tube. Do not mix the yeast solution.

    b. Slowly add 1 mL of NaOH down the side of the tube and mix by tapping the tube with your finger.

    c. Slowly add about 10 drops of Lugol's iodine. Record your results on the data sheet.

        Q7. *What happened in each tube?*

        Q8. *Which tube was the positive control? Which tube was the negative control?*

        Q9. *What do the results of this experiment tell you about the production of alcohol by yeast?*

Formation of Acetyl-CoA

In the second stage of aerobic respiration (cellular respiration in the presence of oxygen), the pyruvate formed by glycolysis enters the mitochondria. Here, each pyruvate molecule is converted into acetyl-CoA, a molecule with two carbon atoms. The other carbon atom is lost as carbon dioxide. Enough of the coenzyme NAD^+ is reduced to generate five molecules of ATP. See Figure 6-6.

    The acetyl-CoA may enter the third stage of cellular respiration, the citric acid cycle, in which more ATP is generated. However, if the cell already has a high level of ATP, it may use the acetyl-CoA to synthesize lipids, thereby storing the energy for later use.

Figure 6-6
Formation of Acetyl-CoA.
Stage 2 of cellular respiration.

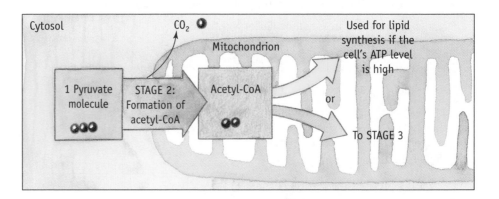

Citric Acid Cycle

The third stage of cellular respiration is the **citric acid cycle,** also called the **Krebs cycle** (see Figure 6-7). The citric acid cycle is a series of reactions in which the carbon atoms in acetyl-CoA are converted into carbon dioxide. In addition, three molecules of $NADH$ and one molecule of $FADH_2$ (a coenzyme like NAD^+) are formed. Because oxygen is the ultimate acceptor of electrons carried by coenzymes, the citric acid cycle is also aerobic. In addition to the ATP formed when $NADH$ and $FADH_2$ are oxidized in the electron transport chain, the citric acid cycle generates two molecules of GTP (guanosine triphosphate), which carries the same amount of energy as ATP. In total, 20 molecules of ATP can be produced from one molecule of glucose in one turn of the citric acid cycle.

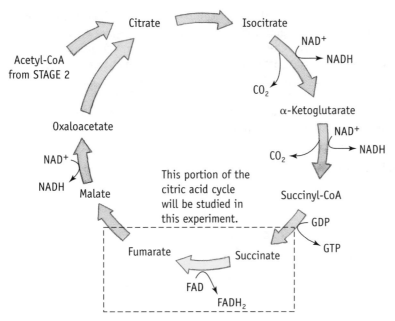

Figure 6-7
Citric Acid Cycle. Stage 3 of cellular respiration.

In this experiment, you will look at one reaction in the citric acid cycle: the conversion of succinate into fumarate (see Figure 6-8). The enzyme that catalyzes this reaction is **succinate dehydrogenase.** As its name suggests, it is involved in the removal of hydrogen atoms and electrons from succinate. The coenzyme FAD works along with the enzyme to accept two hydrogen atoms and their electrons: FAD is reduced to $FADH_2$.

$$
\begin{array}{c}
COO^- \\
| \\
H-C-H \\
| \\
H-C-H \\
| \\
COO^-
\end{array}
\quad
\xrightarrow[\;FAD\quad\quad FADH_2\;]{\text{Succinate dehydrogenase}}
\quad
\begin{array}{c}
COO^- \\
| \\
H-C \\
\parallel \\
H-C \\
| \\
COO^-
\end{array}
$$

Succinate   Fumarate

Figure 6-8
During the Citric Acid Cycle. Succinate is converted into fumarate. Two hydrogen atoms are transferred from succinate to FAD.

Q10. Which molecule is the substrate? Which is the product?

Q11. Is succinate reduced or oxidized?

During the final stage of cellular respiration, $FADH_2$ will enter the electron transport chain, in which, through a series of steps, it gives up its hydrogen atoms and electrons. The hydrogen atoms and electrons eventually react with oxygen to form water.

$$FADH_2 + 1/2\ O_2 \xrightarrow{\text{electron transport chain}} FAD + H_2O$$

Q12. Is $FADH_2$ reduced or oxidized in this reaction? What about the oxygen?

The indicator you will be using, **methylene blue,** can act as an alternate electron acceptor when oxygen is not present. If methylene blue is reduced, it loses its blue color and becomes colorless.

$$
\begin{array}{c}
FADH_2 + \text{methylene blue} \\
\text{(blue)}
\end{array}
\longrightarrow
\begin{array}{c}
FAD + \text{methylene blue} \cdot H_2 \\
\text{(colorless)}
\end{array}
$$

The citric acid cycle, like the other aerobic stages of cellular respiration, occurs in mitochondria. The source of mitochondria you will use is hamburger. Muscles require lots of energy, so muscle cells have many mitochondria. Succinate dehydrogenase, as well as some residual succinate, should be present in hamburger. If succinate is oxidized, FAD is reduced to $FADH_2$. You can test for the presence of $FADH_2$ by adding methylene blue: if the methylene blue is reduced by $FADH_2$, it will lose its color. As a control, you will use **succinic acid,** the acid form of succinate. Succinic acid is easier to maintain in the lab and will produce the same reaction with succinate dehydrogenase that succinate does.

Materials/Equipment

Deionized water in three or four flasks with pipettes

Succinic acid, about 50 mL, in flask with pipette

Malonic acid, about 50 mL, in flask with pipette

Waste beaker

Tub for used test tubes

37°C water bath

Per group of two or three students:
    One dropper bottle with 0.08% methylene blue
    Four test tubes
    Glass stir rod
    Test tube rack
    25 mL hamburger puree in 50 mL beaker
    Wax pencil

1. Label your tubes with your group and "1," "2," "3," and "4."

2. Tube 1

    a. Put 5 mL of deionized water into the tube. Using the level of water in this tube, mark the same level with a wax pencil on the other tubes.

    b. Add 10 drops of succinic acid.

    c. Add three drops of methylene blue.

    d. Mix the contents with a glass stir rod until the color is uniform throughout the mixture.

    e. Place the tube into a 37°C water bath. Note the time this tube was placed in the water bath on the data sheet.

3. Tube 2

    a. Fill the tube to the mark with hamburger puree.

    b. Add 10 drops of succinic acid.

    c. Add three drops of methylene blue.

    d. Mix the contents with a **clean** glass stir rod until the color is uniform throughout the mixture.

    e. Place the tube into a 37°C water bath. Note the time this tube was placed in the water bath on the data sheet.

Malonic acid has a similar structure to succinic acid, so succinate dehydrogenase will bind to it. However, two hydrogen atoms cannot be removed because they are from succinic acid; therefore, no product is formed (see Figure 6-9). Similarly, FAD cannot bind to the hydrogen atoms and their electrons.

4. Tube 3

    a. Fill the tube to the mark with hamburger puree.

    b. Add 10 drops of malonic acid.

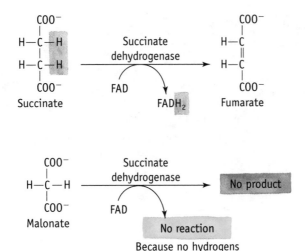

Figure 6-9
Malonate vs. Succinate. Succinate dehydrogenase can bind to succinate and malonate but cannot catalyze a reaction with malonate.

    c. Add three drops of methylene blue.

    d. Mix the contents with a **clean** glass stir rod until the color is uniform throughout the mixture.

    e. Place the tube into a 37°C water bath. Note the time this tube was placed in the water bath on the data sheet.

> *Q13. What do you expect to happen in this tube? How will the results differ from the results of Tube 2? Explain your hypothesis.*

5. Tube 4

    a. Fill the tube to the mark with hamburger puree.

    b. Add 10 drops of water.

    c. Add three drops of methylene blue.

    d. Mix the contents with a **clean** glass stir rod until the color is uniform throughout the mixture.

    e. Place the tube into a 37°C water bath. Note the time this tube was placed in the water bath on the data sheet.

> *Q14. Will there be a reaction in this tube? How will it differ from the results of Tube 2? Explain.*

6. Check the tubes every 10 minutes. Allow the tubes to sit in the water bath about an hour while you continue the lab. Record the time it takes each tube to change color on the data sheet.

> *Q15. In the tubes that reacted, you may have noticed that the top surface of the mixture stayed blue. Why is this—what is the top of the mixture exposed to that would keep the methylene blue from reacting with the $FADH_2$?*

> *Q16. Was there a reaction in Tube 3? How would it be possible for a reaction to occur in Tube 3? If there was a reaction in Tube 3, how did it differ from the reaction in Tube 2, and why?*

> *Q17. Was there a reaction in Tube 4? Did you expect one? How did it differ from the reaction in Tube 2, and why?*

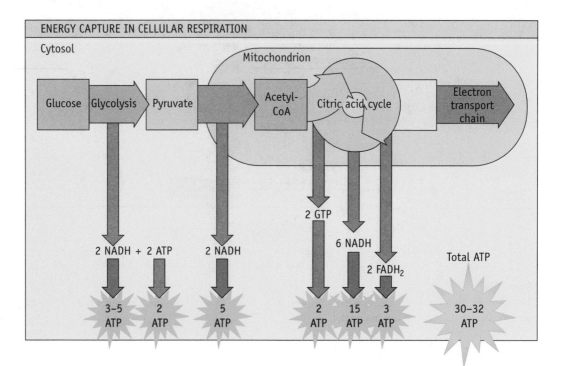

Figure 6-10
Energy Capture in Cellular Respiration. For each glucose molecule, glycolysis produces 2 ATP and 2 NADH; the production of acetyl-CoA yields an additional 2 NADH. The citric acid cycle then yields 2 GTP, 6 NADH, and 2 FADH$_2$. The electron transport chain captures the energy of NADH and FADH$_2$ in 26–28 ATP molecules.

Electron Transport Chain: Oxidative Phosphorylation

The fourth stage of cellular respiration, the electron transport chain, is the stage in which ATP is produced (see Figure 6-10). The FADH$_2$ and NADH generated in the first three stages transfer their hydrogen atoms (and associated electrons) to oxygen through a series of steps. Oxygen is the most electron-hungry atom in the environment, so this is a "downhill" energy-releasing process, much like the movement of a substance into a cell along a concentration gradient. The mitochondria take advantage of the energy in the attraction between the electrons and oxygen and couple it with the "uphill" process of synthesizing ATP. This process is detailed in "How Do Cells Harvest the Energy of Electron Transport?" in the text.

When you are finished with this lab

- Dispose of the test tube contents as instructed
- Remove labels
- Place empty test tubes and stir rods into a wash bin
- Wash your hands thoroughly

Lab 7   Photosynthesis
This lab accompanies Chapter 7 of *Asking About Life.*

Objectives

1. Understand the process by which photosynthetic organisms capture energy from the Sun and convert it into chemical energy that can be used by cells for work

2. Understand the importance of photosynthesis for life on Earth

3. Compare and test for photosynthesis and respiration in an aquatic plant

4. Observe oxygen and starch production in plants

5. Manipulate factors that affect the rate of photosynthesis

6. Observe some structures involved in photosynthesis

Introduction

All the energy you get from the food you eat originally came from the Sun. Remember the first law of thermodynamics? Energy can be neither created nor destroyed; it can only be changed from one form to another. **Photosynthesis** is the process by which most autotrophic organisms change light energy into chemical energy. Plants and other autotrophs (some bacteria and protists) take the energy from sunlight and convert it into the electrical energy of high-energy electrons. This electrical energy is converted into the high-energy chemical bonds of ATP and NADPH (a coenzyme, similar to the NAD$^+$ used in cellular respiration). Finally, it is converted into the chemical bonds in glucose, a stable form in which to store energy. Starch and other carbohydrates can be synthesized from the glucose; the addition of nitrogen, phosphorus, and other minerals from the environment allows proteins, lipids, and amino acids to be produced. Heterotrophs (such as animals and fungi) obtain energy by direct consumption of autotrophs and/or by consumption of other heterotrophs that obtained their energy by consuming autotrophs.

The overall chemical transformation that takes place during photosynthesis can be summarized as follows:

$$6 \text{ CO}_2 + 6 \text{ H}_2\text{O} \xrightarrow{\text{light}} \text{C}_6\text{H}_{12}\text{O}_6 + 6 \text{ O}_2$$

$$\text{carbon dioxide} \quad \text{water} \quad\quad \text{glucose} \quad \text{oxygen}$$

Notice that this reaction is the exact opposite of **cellular respiration**, in which glucose is broken down in the presence of oxygen to produce carbon dioxide and water. Like respiration, photosynthesis involves many chemical reactions and electron transfers. See Chapter 7 in the text for descriptions of these processes.

Photosynthesis takes place in **chloroplasts**, green plastids that contain the pigment **chlorophyll**. (See Figure 7-16 in the text for the anatomy of a chloroplast.)

There are photosynthetic members in the kingdoms Plantae, Eubacteria, and Protista. Examples from each kingdom are shown in Figure 7-1.

 Note: During this lab, you will have some free time—for instance, after you set up the first experiment and while you wait for oxygen production in the second experiment. Spend this time completing as many "Short Observations" exercises (at the end of the main exercises) as possible.

Skip Moody/Dembinsky Photo Associates

A. *Trillium,* a flowering plant in the kingdom Plantae.

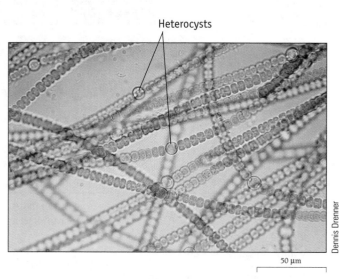

Heterocysts

Dennis Drenner

50 µm

B. *Anabaena,* a cyanobacterium in the kingdom Eubacteria.

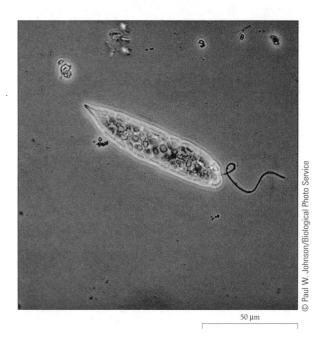

© Paul W. Johnson/Biological Photo Service

50 µm

C. *Euglena,* a unicellular, motile organism in the kingdom Protista.

J.M. Kingsbury

D. *Ulva,* or sea lettuce, a multicellular green algae in the kingdom Protista.

Figure 7-1
Photosynthetic Organisms from Different Kingdoms.

Comparison of Photosynthesis and Respiration

Plants use photosynthesis to produce and store energy, but they must also use energy to perform cellular processes and to grow, just like any other living organism. Like other organisms, plants break down organic molecules such as glucose by cellular respiration. Although photosynthesis can only occur in the presence of light, cellular respiration is always occurring in plants.

In this experiment, you will look at photosynthesis and respiration in the aquatic plant *Elodea.* The plant uses carbon dioxide from the water as it photosynthesizes, and it gives off carbon dioxide as it undergoes cellular respiration. Carbon dioxide is soluble in water, forming **carbonic acid** as it dissolves: the more carbon dioxide in a solution, the more acidic the solution.

$$CO_2 + H_2O \longleftrightarrow H_2CO_3 \longleftrightarrow H^+ + HCO_3^-$$
$$\text{Carbonic acid}$$

This reaction is **reversible:** as carbon dioxide is removed from the solution (during photosynthesis), the solution becomes more basic.

Q1. How would respiration affect the pH of the solution?

You will use the indicator **phenol red,** which changes color with pH as shown in Figure 7-2. In a basic solution, phenol red is pink; in an acidic solution, the phenol red is yellow.

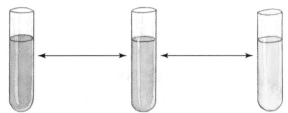

Figure 7-2   Basic pH   Neutral pH   Acidic pH
Phenol Red. This indicator changes color in response to changes in pH.

Materials/Equipment

Three or four 125 mL flasks of phenol red with graduated pipettes

Three or four lights, 200 W

Glass bricks or water-filled beakers for heat traps

Per group of two or three students:
    Two sections of *Elodea,* 10–15 cm
    Four large test tubes, 18×160 mm
    Two 250 mL beakers, to hold test tubes
    400 mL beaker
    Straw
    Forceps

1. Label the four test tubes "1," "2," "3," and "4."
2. Fill the 400 mL beaker to the 150 mL mark with tap water.
3. Add 3 mL of phenol red. The water should be pink, indicating that it is slightly basic.
4. Turn the water more acidic by blowing gently through a straw until the water is orange-pink, halfway between acid (yellow) and basic (red).

   Q2. Why does blowing into the solution cause it to become more acidic?

5. Fill the four test tubes about two-thirds full with this water.
6. Use forceps to place an *Elodea* sprig into Tubes 1 and 3, cut end up. Make sure the sprigs are completely immersed in the water.
7. Place Tubes 1 and 2 into a 250 mL beaker near the light and protected by the heat trap.
8. Place Tubes 3 and 4 into a 250 mL beaker, as shown in Figure 7-3. Place the beaker in the dark (for example, in a cupboard or other space indicated by your

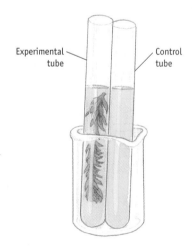

Experimental tube   Control tube

Figure 7-3
Setup for Respiration/
Photosynthesis Experiment.

instructor). Let the reactions proceed for about an hour while you continue with the next lab exercise.

Q3. What do you predict will happen to the color of the solution in each tube?

9. In about an hour, compare the color changes in the experimental tubes with the color in the control tubes. Lifting the plant slightly out of the water with forceps and holding the tubes in front of a white piece of paper will help you see the color. Record your results on the data sheet.

Q4. Were your predictions correct? If the results differed from your predictions, explain why they differed.

Q5. How would the results in Tube 1 differ if you could keep the Elodea from respiring? Explain your answer.

Q6. Cellular respiration and other processes that break down energy-rich molecules (for example, the burning of gasoline) give off CO_2. How does the photosynthetic activity of plants help to balance the effects of these processes?

Light Intensity and Photosynthesis

Many factors influence plant productivity. Photosynthesis requires light, carbon dioxide, and water. Anything that limits the availability of these components will limit the rate of photosynthesis. The length of the day, the angle at which sunlight enters the atmosphere, pollution, and shade all limit the amount of sunlight that reaches the surface of a leaf.

This experiment allows you to indirectly measure the amount of photosynthesis that takes place in different intensities of light by placing a photosynthesizing plant different distances from a light source.

Materials/Equipment

Three or four lights, 150–200 W

Glass bricks or water-filled beakers for heat traps

Meter stick for measuring distance

Per group of two or three students:
  Two photosynthesis chambers:
      125 mL flask
      Two-hole stopper to fit in flask (size 4)
      Pasteur pipette with a marking about 1.5 cm from the stopper
      3 cc syringe
  One sprig of *Elodea*, 15 cm
  0.1 M NaHCO$_3$ (baking soda), about 400 mL
  Small metric ruler

 The lamps used in this experiment are very hot. Use caution when working around them.

1. Into one photosynthesis chamber, place a sprig of *Elodea,* cut end up.
2. Fill both chambers to the top with the 0.1 M NaHCO$_3$ solution. (NaHCO$_3$, or baking soda, is a source of carbon dioxide.)
3. Push the plunger of the syringe all the way down, then insert the stopper with the pipette and syringe into the photosynthesis chamber, as shown in

Figure 7-4. **Syringes are very sharp.** If there are air bubbles or spaces under the stopper, remove and dry the stopper, fill the chamber again, and reinsert the stopper. A little solution will move into the pipette; adjust this level to the mark on the pipette by raising or lowering the plunger in the syringe. If the stopper will not stay in the flask, it may be because the $NaHCO_3$ makes the solution a little slippery. Try removing the stopper and rinsing, drying, and reinserting it.

Figure 7-4
Setup for Light
Intensity/Photosynthesis
Experiment.

4. Trial 1: 25 cm

    a. Place both chambers 25 cm from the light source, with a heat filter between the lamp and chambers. **Caution: These lights are hot.** Make sure both chambers are exposed equally to the light.

    b. Allow five minutes for equilibration. Readjust the level of the solution so that it is at the zero mark on the pipette.

    c. After equilibration, start the experiment: Measure the level of solution in the pipettes (from the starting mark) every 3 minutes for 15 minutes. Record the values on the data sheet. To obtain a "true" reading from the experimental chamber, the value measured from the control chamber must be subtracted from the value measured from the experimental chamber. Why is this?

    Q7. What is happening to the level of solution in the experimental chamber?

    Q8. What is causing this level of solution to change?

    Q9. What do you expect will happen if the chambers are placed farther from the light?

5. Trial 2: 75 cm

    Now, place both chambers 75 cm from the light source, with the heat filter between the lamp and chambers. Allow 5 minutes for equilibration, readjust the level of solution in the pipettes to the starting mark, and repeat the experiment. Record the readings on the data sheet.

6. After calculating the true values obtained from the experimental chambers, graph your results on the data sheet. Use different colors (or different types of lines) to distinguish between the trials.

7. If time allows, design another experiment to test the effects of light intensity on photosynthesis. For instance, you could look at the effects of shade, the effects of a different distance from the light source, or the effects of placing the chambers in the dark. Alternatively, try the "Differential Light Absorption for Photosynthesis" experiment under "Short Observations."

    a. Write the hypothesis you are testing and how you will perform the experiment.

    b. Make sure to equilibrate the chambers before you begin.

    c. Take readings of the level of solution in the pipettes every 3 minutes for 15 minutes in the new environment.

    d. Was your hypothesis correct? Compare your results with the results from the first two trials.

    Q10. What is the effect of light intensity on the photosynthetic rate? What is your evidence for your answer?

    Q11. What is the role of the $NaHCO_3$ in the tubes? What would happen if you used plain water instead?

    Q12. What component (or the lack of what component) in the solution would eventually limit photosynthesis by the Elodea in the flask?

    *Q13. In this experiment, you were measuring oxygen production with a linear measurement. What are two ways you could determine the **volume** of oxygen produced during this experiment? In what units would this measurement be?*

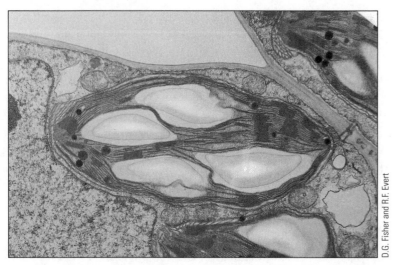

Figure 7-5
Electron Micrograph of Three Chloroplasts. The chloroplast in the center of the micrograph contains four large, white starch grains.

Starch Production during Photosynthesis

Although glucose is commonly thought of as the end product of photosynthesis, there is little free glucose in photosynthesizing cells. Instead, the glucose is converted into either sucrose, the major sugar transported between cells in plants, or starch, the major storage carbohydrate in plants. In fact, during intense periods of photosynthesis, some products of photosynthesis are temporarily stored right in the chloroplast in the form of starch, as shown in Figure 7-5.

In this experiment, you will identify the areas in a leaf that are undergoing photosynthesis by preparing the leaf and staining it with Lugol's iodine. Recall that iodine reacts with starch to turn from brownish yellow to bluish black. You will use a variegated leaf (one that has areas of different colors).

> Q14. If your leaf has areas that are white, red, and green, which areas do you suppose will darken when the indicator is added? What is your explanation?

Materials/Equipment

Coleus plant that has been actively photosynthesizing

Four or five boiling water baths:
   Hot plates
   400 mL beakers with water
   Boiling chips

95% ethanol, four or five 400 mL beakers in 60°C water bath under hood

Four or five glass Petri dishes half full of Lugol's iodine

Per group of two or three students:
   Forceps
   Petri dish
   Paper towel

1. Obtain a *Coleus* leaf that has two or three colors. Draw the pattern of the colors on the data sheet, labeling the colors.

2. Place the leaf into a boiling water bath for about a minute. Keep track of your leaf so that it does not get mixed up with another group's leaf. Carefully remove the leaf with forceps and place it into your Petri dish.

    Q15. What happened to your leaf in the boiling water?

    Q16. What does this tell you about the solubility of pigments in water? Is chlorophyll soluble in water?

3. Transfer your leaf to the hot alcohol bath under the hood. Keep track of your leaf so that it does not get mixed up with another group's leaf. Leave it for 3 or 4 minutes, until all pigment has leached out. Remove the leaf and place it back into the Petri dish.

4. Transfer the whitened leaf to a Petri dish filled with Lugol's iodine. When it has darkened, remove the leaf and blot it gently with a paper towel.

5. Compare the stained pattern with the original pattern you drew.

    Q17. How did your prediction compare with your results? Explain the cause of any differences.

    Q18. What pigment was most actively photosynthetic under the growing conditions of the Coleus? How might the results have been different if the plant had been growing in green light?

    Q19. Variegated plants (plants with white streaks or spots) do not make good weeds. They are cultivated for their attractive appearance but often do not fare well in nature. Why might this be, based on your results from this test?

Short Observations

Observing Stomata

Materials/Equipment

> *Zebrina* plant
>
> Compound microscopes
>
> Clean glass slides
>
> Coverslips
>
> Deionized water in flasks with pipettes
>
> Container for used slides
>
> Sharps disposal
>
> Lens paper

In the late 17th century, scientists in England and Italy discovered tiny pores called **stomata** (*stoma* = mouth) in the leaves of plants. The stomata appeared to let air into the leaves, so this discovery was the first clue that plants interacted with the air somehow. See the electron micrograph of stomata in Figure 7-2 in the text.

> *Q20. What are two components of air that are essential for plants to live? What biological processes are each of these components used for?*

Stomata can be easily observed on a small piece (slightly less than 1 cm²) of leaf from the plant *Zebrina*. Put the piece upside down on your slide in a drop of water, add a coverslip, and observe it under a microscope. Find the stomata: Each is made of two cells with an opening into the interior of the leaf between them. They can open and close depending on the needs of the plant.

Q21. Aquatic plants have stomata on both the top and bottom surfaces of their leaves; terrestrial plants tend to have more stomata on the bottom surfaces of their leaves. Both types of plants must exchange gases with the environment, so why do terrestrial plants have fewer stomata on the tops of their leaves? What function does this placement of stomata accomplish?

What Is Light?

Materials

Prisms or crystals for refracting light

Sheet of white paper

Light is a form of electromagnetic radiation, a form of energy with properties both "wavelike" and "particle-like" (see Figures 7-8 and 7-9 in the text). The light that we see every day is **visible light,** a mixture of different wavelengths of light. Different wavelengths of visible light are seen as different colors: 400 nanometer (nm) light looks violet; 500 nm light looks blue-green. Light that appears white is made of a mixture of many wavelengths of light. A **prism** (a transparent piece of plastic or glass that has two nonparallel faces) bends, or **refracts,** light waves. The longer the wavelength, the less the light wave is bent. When white light is sent through a prism, as shown in Figure 7-6, it will be separated into its component wavelengths by the prism, allowing you to see the different colors.

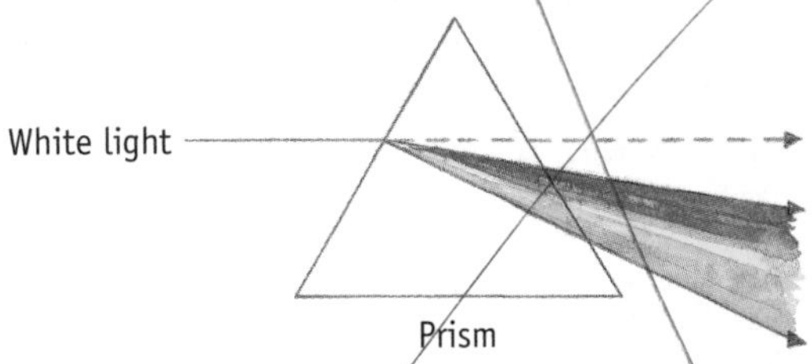

Figure 7-6
Refraction. A prism separates white light into its component wavelengths, which we see as different colors of light.

White light

Prism

Use one of the prisms provided to refract white light into its components: hold the prism between a bright artificial light source or the Sun and a piece of white paper. You may need to experiment with the angle at which you hold the prism before the colors will shine on the paper.

You have probably seen this phenomenon occur along the edges of a piece of glass or a mirror. A rainbow is formed in the same way: sunlight is broken into its component colors by drops of water, which act as prisms.

Q22. What colors of light do you see?

Q23. If you look at different lights (such as sunlight and lamplight), do they refract into different colors?

Differential Light Absorption for Photosynthesis

Materials/Equipment

Those used for the "Light Intensity and Photosynthesis" experiment

Colored filters (red, blue, and green)

Visible light is a mixture of different wavelengths of light, each of which appears as a different color. An object that appears blue absorbs all the wavelengths but blue; the blue wavelength of light is reflected, and that is the color you see. With that in mind, do you think chloroplasts absorb green light?

Examine Figure 7-7. It illustrates an **action spectrum,** which demonstrates the effectiveness of different wavelengths of light for photosynthesis. The **absorption spectrum** for chlorophyll *a,* the major photosynthetic pigment, is in green. The absorption spectrum shows the colors of light absorbed by chlorophyll *a.*

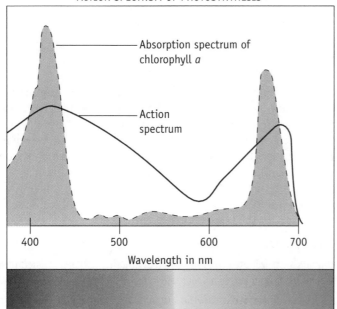

ACTION SPECTRUM OF PHOTOSYNTHESIS

Absorption spectrum of chlorophyll *a*

Action spectrum

400   500   600   700

Wavelength in nm

Figure 7-7
Action Spectrum. An action spectrum is the range of wavelengths that effectively promotes a light-dependent process. The action spectrum of photosynthesis, for example, roughly parallels the absorption spectrum of the pigment chlorophyll *a*.

You can test the colors of light *Elodea* can absorb and use for photosynthesis by setting up the photosynthesis chambers used in the "Light Intensity and Photosynthesis" experiment. Wrap the experimental chamber with a sheet of colored plastic that will filter out other colors of light.

Form a hypothesis using the materials and the colors of filters available, and design an experiment to test the hypothesis. Make sure to include a control.

1. Write down the hypothesis you are testing and how you will perform the experiment.

2. Change the NaHCO₃ solutions in the photosynthesis chambers so that they are fresh. Follow the instructions in the light intensity experiment for using the chambers.

3. Make sure to equilibrate the chambers before you begin.

4. Take readings of the level of solution in the pipettes every 3 minutes for 15 minutes. Make a table to record your measurements.

Q24. *What were the results of your experiment? Did they support your hypothesis? Explain why they did or did not support your hypothesis.*

Q25. *How would you redesign your experiment if you were to repeat it? What is your reasoning?*

Q26. *Incandescent light bulbs (typical light bulbs) emit light in the yellow–orange range; fluorescent bulbs (tube-shaped bulbs often used to light schools and other buildings) emit light in the blue–violet range. Nurseries and hardware stores sell "grow lights" specifically designed for growing plants. What color or colors of light do you suppose these lights emit? Explain your answer.*

Observing Chloroplasts

Materials/Equipment

Elodea

Compound microscopes

Clean glass slides

Coverslips

Deionized water in flasks with pipettes

Container for used slides

Sharps disposal

Lens paper

Much energy from sunlight is "lost" as heat. Some molecules, **pigments,** can absorb visible light and convert it to the electrical energy of high-energy electrons, thereby beginning the process of photosynthesis. One such photosynthetic pigment, chlorophyll, gives plants their green color. Chlorophyll absorbs red and blue wavelengths of light and lets the green wavelengths pass through or be reflected—which is how we see the green. Chlorophyll is contained in special organelles called chloroplasts.

You may recall observing chloroplasts in *Elodea* in Lab 3. Make a wet mount of one leaf and look at it under a compound microscope. The cell walls are fairly distinct and form the structure for the box-shaped cells.

> Q27. *How can you identify the chloroplasts?*

> Q28. *Carotenoids are yellow or orange photosynthetic pigments. Which of the following wavelengths of light do the carotenoids absorb: red, blue, yellow, and/or green?*

> Q29. *Few plants can survive as **albinos** (a genetic mutation in which the affected individual produces no pigment). One of the few is the California redwood: Sometimes an albino tree will grow as a shoot from a normal tree. It shares the root system of the normal tree and will never grow very large. Where does such a tree obtain its energy?*

Observing Leaf Sections

Materials/Equipment

Prepared slides of typical leaf cross-section

Compound microscopes

Lens paper

Take a prepared slide of a typical leaf cross-section and observe it under a microscope. When it is under a magnification at which you can easily see the different cells in the leaf (100X or 400X), look for the following cells and tissues:

- **Epidermis:** a layer of cells covering the top and bottom surfaces. This layer is protective; sometimes it is coated with a thin layer of a waxy substance (which stains pink) that protects the leaf from water loss.

> Q30. *Look for stomata in the epidermis. Do you expect to find more stomata on the top or the bottom surface of the leaf?*

- **Palisade parenchyma:** a layer of primarily photosynthetic cells. They are packed together in an organized fashion and filled with chloroplasts.

> Q31. *Would these cells be at the top or the bottom surface of the leaf? Explain.*

- **Spongy parenchyma:** a type of tissue made of cells that are shaped irregularly and are used primarily for storage and transport.

> Q32. *How can you tell that these cells perform less photosynthesis than the palisade parenchyma cells?*

You may also see bundles of small, regularly shaped cells, which are the **vascular bundles** of the leaf. These are the "veins" of the plant; they transport sugars, water, and minerals throughout the plant.

Draw the leaf section, labeling the cell types and their functions.

Lab 8

Mitosis and Meiosis
This lab accompanies Chapters 8 and 9 of *Asking About Life.*

Materials/Equipment

Pop beads, about 600 each of two colors (such as red and yellow)

Magnetic "centromeres," about 120

Compound microscopes

Immersion oil (if appropriate)

Lens paper

Coverslips

Clean glass slides

Container for used slides

Sharps disposal

Prepared slides:

   Whitefish blastula (mitosis)

   Onion (*Allium*) root tip (mitosis)

Prepared leek root tips that are hydrolyzed in water, with forceps

1% acetocarmine, three or four bottles with droppers

Matches

Razor blades

Alcohol lamps or the equivalent (emergency candles or tea lights)

Forceps

Probes

Clothespins for holding slides while heating

Objectives

1. Understand the importance of mitosis and meiosis

2. Understand the mechanics of mitosis and meiosis by making models of the stages

3. Observe the stages of mitosis in plant and animal cells

4. Compare mitosis and meiosis in terms of the production of genetic diversity

Introduction

All 10 trillion cells in your body originated from a single cell, the zygote that was the result of the fertilization of an egg by a sperm. Half the genetic information in the nucleus of that zygote was from your mother; half was from your father. All this information was passed to each of your 10 trillion cells.

Every minute, your body produces about 300 million new cells. Your hair and nails are growing; your blood cells and skin cells are being replaced; injuries are healing; and so on. To go from one cell to many cells, to maintain a multicellular body, and to allow reproduction, cells must be able to accurately duplicate themselves. The pro-

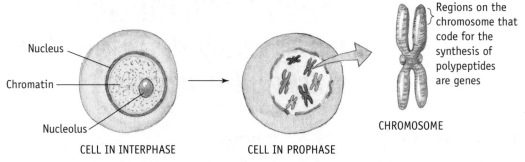

Nucleus

Chromatin

Nucleolus

CELL IN INTERPHASE   CELL IN PROPHASE

Regions on the chromosome that code for the synthesis of polypeptides are genes

CHROMOSOME

cess by which cells divide is **mitosis**; the process by which sexually reproducing animals generate **gametes** (eggs and sperm, in most animals) is **meiosis**.

First, it is important to understand a little about chromosomes. The nucleus of a cell contains **nucleoli**, in which ribosomal RNA is made, and **chromatin**, a combination of DNA and proteins. (See how the proteins and DNA are combined to form chromosomes in Figure 8-7 in the text.) The chromatin is organized into **chromosomes**, as shown in Figure 8-1. The number of chromosomes in the nucleus depends on the species—for example, fruit flies have 8, and humans have 46. Each chromosome can be divided into regions, called **genes**. Each gene carries the genetic information to produce a particular protein or to help to regulate the expression of other genes. For instance, one gene might influence eye color, another might influence hair color, and yet another might have the genetic information to produce the protein amylase.

Mitosis

The cell cycle consists of three major phases, of which mitosis is one (see Figure 8-2).

- **Mitosis** is the process by which unicellular organisms reproduce asexually and by which multicellular organisms grow and repair themselves. It is the part of the cell cycle in which the nucleus divides.
- **Cytokinesis** is the division of the cytoplasm and the formation of two plasma membranes to form two new cells.
- **Interphase** is the time between dividing. This is when growth, replication of DNA, and preparation for mitosis occur.

Figure 8-2
Cell Cycle.

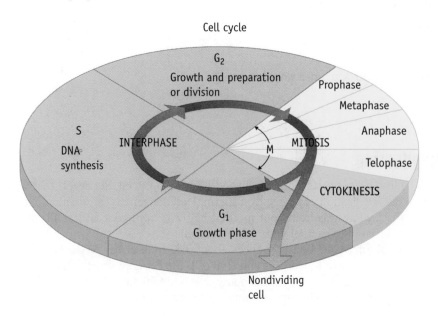

Cell cycle

G_2
Growth and preparation or division

Prophase

Metaphase

Anaphase

Telophase

CYTOKINESIS

S
DNA synthesis

INTERPHASE

M

MITOSIS

G_1
Growth phase

Nondividing cell

Pop-Bead Mitosis

Many processes studied by scientists are difficult or impossible to observe directly and are understood by the use of models. For instance, many astronomical events take place over millions of years and can only be observed by simulation, usually by a computer model. Chemists use molecular models to understand the structure of molecules too small to be directly observed.

To track the course of each chromosome during mitosis, you will simulate mitosis using pop beads (red and yellow, or two other colors). Your organism will have two pairs of chromosomes. (Humans have 23 pairs, so they have a total of 46 chromosomes.) Make the two pairs of chromosomes as illustrated in Figure 8-3. One pair should be short: make one chromosome with red beads and the other chromosome with yellow beads; each chromosome should have the same number of beads. The other pair should be long: one red and one yellow chromosome, both with the same number of beads. The two short chromosomes are **homologous chromosomes** (*homo* = same): they have genes for the same traits in the same order, but they are not identical. For example, the eye color gene on one chromosome may code for blue eyes; the homologous chromosome's eye color gene may code for brown eyes. A cell with two sets of chromosomes is described as **diploid**.

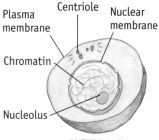

Plasma membrane — Centriole — Nuclear membrane
Chromatin
Nucleolus

INTERPHASE

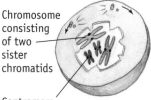

Chromosome consisting of two sister chromatids
Centromere

PROPHASE

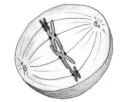

METAPHASE

ANAPHASE

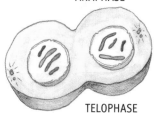

TELOPHASE

**Figure 8-4
Stages of Mitosis.**

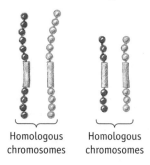

Homologous chromosomes   Homologous chromosomes

**Figure 8-3
Make Two Pairs of Homologous Chromosomes.**

Interphase

During interphase, the nucleus and nucleoli are intact, and the DNA within the nucleus is indistinct—the chromatin (DNA and associated proteins) in the nucleus appears granular. In this phase, the amount of DNA in the cell is doubled.

1. Replicate your chromosomes by making an identical set of each pop-bead chromosome.

2. Connect the two identical **sister chromatids** by the magnetic **centromere**, as shown in Figure 8-5.

Your cell is ready to begin mitosis. Mitosis is divided into four phases. See Figure 8-4.

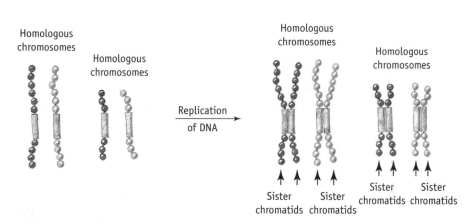

Homologous chromosomes

Homologous chromosomes

Replication of DNA →

Homologous chromosomes

Homologous chromosomes

Sister chromatids   Sister chromatids   Sister chromatids   Sister chromatids

**Figure 8-5
Interphase.** The cell's DNA replicates so that each chromosome consists of two sister chromatids.

Prophase

During **prophase,** the chromatin starts to condense into discrete chromosomes, and the chromatids are visible. The nuclear membrane breaks down, and a **mitotic spindle** develops. In animal cells, this consists of two pairs of tiny **centrioles** surrounded by a band of hollow tubes called **microtubules.** Each chromatid develops a **kinetochore** that helps attach the mitotic spindle to the centromere. See Figure 8-6. The microtubules attached to the kinetochores start to move the chromosomes to the center of the cell.

Figure 8-6
How a Pop-Bead Chromosome Corresponds to a Chromosome in a Cell.

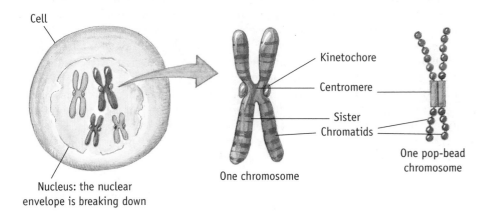

Cell

Nucleus: the nuclear envelope is breaking down

Kinetochore

Centromere

Sister Chromatids

One chromosome

One pop-bead chromosome

Metaphase

By the end of **metaphase,** all of the chromosomes have moved to the center of the cell, halfway between the two poles of the mitotic spindle. Place your chromosomes into their positions during metaphase. Draw your chromosomes, adding the mitotic spindle, centrioles, and kinetochores. Label each structure.

Anaphase

During **anaphase,** the sister chromatids are separated from each other. The centromeres split, and the chromatids are pulled in opposite directions by the microtubules that attach at the kinetochore. As the microtubules pull on the centromeres, the rest of each chromatid trails along as if it is being dragged. Put your chromosomes into their positions in anaphase. Draw them, adding the mitotic spindle, centrioles, and kinetochore. Label each structure.

Telophase

Telophase is the last phase of mitosis. The sister chromatids have been completely separated; the mitotic spindle, centrioles, and kinetochore disperse; and the chromatin assumes its indistinct, grainy appearance. Nuclear membranes form around the two separate nuclei. Then cytokinesis occurs, in which the cytoplasm divides and two separate plasma membranes form. The result is two daughter cells. Rearrange your chromosomes into their telophase positions. Draw them, adding the nuclear membrane and the plasma membrane of each cell.

> Q1. *How many chromosomes does each daughter cell have? How many did the parent cell have? Is this number the same or different?*

> Q2. *The parent cell was diploid: it had two sets of chromosomes. Are the daughter cells also diploid?*

Viewing Mitosis in Real Cells

If you refer back to the diagram of the cell cycle (see Figure 8-2), you will notice that for most cells, interphase is the longest phase. Most of the cells you see today will be in interphase: the chromatin will appear grainy and indistinct inside the nucleus. You will not be able to tell if the DNA has replicated. Nucleoli will be visible.

When a cell enters prophase, the chromosomes begin to condense, becoming visible as darkly stained structures. They will appear this way until the end of telophase. Nuclear membranes will form, and the chromatin will appear indistinct again. As you look for different stages of mitosis, remember that it is a continuous process: you may see cells that are between stages.

Mitosis in Animal Cells: Whitefish Blastula

The **blastula** is a stage in animal development that occurs shortly after fertilization. Cells are dividing rapidly to form a hollow ball of cells from the original fertilized egg. This is a prepared slide, so the specimen was preserved at this stage of development. Refer to Figure 8-6 in the text.

1. Starting at the lowest power, locate and focus on the specimen on the slide. Work your way to the high power lens (40X), looking for cells in each stage of mitosis. You should be able to see the microtubules of the mitotic spindle as pink fibers.

2. If your microscope has an oil immersion lens, you may want to look at the cells at this power. Refer to Lab 3 if you do not remember how to use oil immersion. At this power, you may be able to see the centrioles at the poles of the mitotic spindle.

3. Draw a cell in each stage of mitosis. Label the stage and describe what is happening in the cell. Note that telophase will be accompanied by the formation of a **cleavage furrow,** a groove that forms around the cell as the plasma membrane is pinched to divide into two cells. See Figure 8-9 in the text.

Mitosis in Plant Cells: Leek Root Tip

Your instructor may want you to substitute this preparation with the next exercise, which uses a prepared slide. Wait for instructions before proceeding.

Cells in the tip of a root divide rapidly as the root grows and pushes through the soil. Leeks from the grocery store were put into water until the roots grew; these roots were cut and preserved for your use. Because they are from a fresh specimen, you must stain the cells to see the stages of mitosis occurring.

Materials/Equipment

One coverslip

One clean glass slide

One prepared leek root tip, hydrolyzed

Acetocarmine stain

Matches

Razor blade

Alcohol lamp

Forceps

Two probes

Clothespin for holding the slide while heating

1. Place one root tip onto a slide with a pair of forceps. See Figure 8-7 for steps 1–6.
2. Use a razor blade to cut and discard all but 3–4 mm of the whitish, pointed tip of the root.
3. Use the two probes to gently tease the tissue into small pieces.
4. Add one or two drops of acetocarmine stain to the tissue on the slide.
5. Attach a clothespin to one end of the slide, and heat the stain and sample by waving the central portion of the slide about an inch above the flame for about 5 seconds. Do not let the stain boil off or the slide dry out. You may add more stain if it starts to dry too quickly. If the stain does boil off, start over with a new root tip.

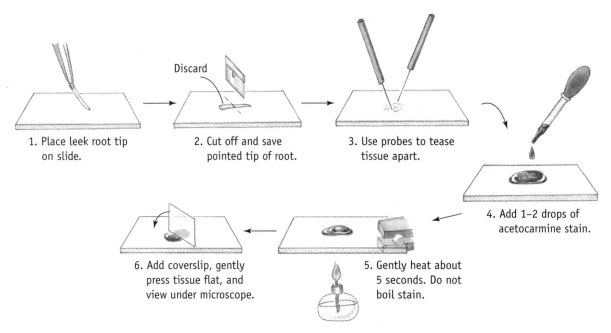

1. Place leek root tip on slide.

Discard

2. Cut off and save pointed tip of root.

3. Use probes to tease tissue apart.

4. Add 1–2 drops of acetocarmine stain.

5. Gently heat about 5 seconds. Do not boil stain.

6. Add coverslip, gently press tissue flat, and view under microscope.

Figure 8-7
Preparing a Leek Root Tip Slide, to Show Mitosis.

6. Remove your slide from the heat and add a coverslip. Using your finger or the blunt end of a probe, **gently** press the tissue flat.

7. Examine your slide under a compound microscope. Try to find one example of each phase of mitosis. Draw each stage. Label each drawing and describe what is happening in the cell.

> Q3. *How do the leek cells compare with the whitefish cells in size?*

> Q4. *What did the acetocarmine stain bind to in the leek cells (what was stained pink)?*

Mitosis in Plants: Onion Root Tip

The onion root tip slide is of a thin section through the tip of a root.

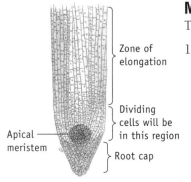

Zone of elongation

Dividing cells will be in this region

Apical meristem

Root cap

Figure 8-8
Regions of an Onion Root Tip.

1. As you focus using low and medium power, notice the three regions of the root tip, shown in Figure 8-8.
   - **Root cap:** at the pointed tip of the root. The root cap is composed of cells that produce a polysaccharide slime, which lubricates the path of the growing root in the soil.
   - **Apical meristem:** directly behind the root cap. In this region, the cells are undergoing cell division; you should look here for cells in the stages of mitosis.
   - **Elongation zone:** above the apical meristem. In this zone, the cells are growing longer and differentiating into the tissues of a mature root.

2. Locate the region of cell division near the apical meristem to find the stages of mitosis. Work up to high power. Find one example of a cell in each phase of mitosis.

3. Draw each stage. Label each drawing, and describe what is happening in the cell. In cells undergoing telophase, note the formation of a new cell wall between newly formed daughter cells. While this new cell wall is being formed, it is called a **cell plate**. Small channels, called **plasmodesmata**, will be left in the new cell wall. Figure 8-10 in the text shows the formation of a new cell wall.

> Q5. *What are these channels for?*

> Q6. *Why is the cell wall between plant cells not solid?*

> Q7. *How do the onion cells compare with the whitefish cells in size?*

Meiosis

Meiosis is a type of cell division used by sexually reproducing, multicellular organisms for reproduction. How does this compare to reproduction by mitosis? Mitosis is an easy way for unicellular organisms to reproduce. The budding yeast cells in Lab 3 were reproducing by mitosis. Some multicellular organisms can also reproduce using mitosis by splitting off a single cell or group of cells that grows into a new, multicellular organism. Reproduction by mitosis is called **asexual reproduction,** and it results in diploid offspring genetically identical to the diploid parent that created them.

Sexual reproduction, however, results in offspring that have genetic information from two parents, creating a new combination of genetic characteristics. To accomplish this, each diploid parent must be able to form **haploid** cells (with one set of chromosomes, instead of two). These haploid cells, called gametes, can then combine to form a new, diploid individual. **Meiosis** is the process by which sexually reproducing animals produce haploid gametes, as shown in Figure 8-9.

The plant life cycle is different: it is divided into a diploid stage and a haploid stage, as shown in Figure 8-10. The diploid stage forms haploid spores by meiosis. Each spore grows into a haploid **gametophyte** that will produce gametes. These gametes will then combine to form a new, diploid stage of the plant.

Q8. If the gametophyte is haploid, what process does it use to form gametes?

In many plants and animals, the female gamete is an **egg** or **ovum,** and the male gamete is a **sperm** or **spermatozoan.**

Using the same four chromosomes with which you began, you can simulate meiosis for the same hypothetical organism.

During meiosis, there are two cell divisions, called **meiosis I** and **meiosis II.** You may refer to Figure 9-9 in the text as you conduct your simulation.

Meiosis I
DNA Replication
Before prophase I of meiosis, the DNA replicates itself so that each chromosome consists of two sister chromatids, as in mitosis. Replicate your chromosomes and connect the sister chromatids to each other at the centromere.

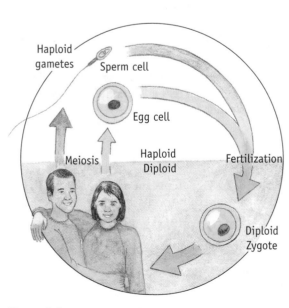

Figure 8-9
Typical Animal Life Cycle.

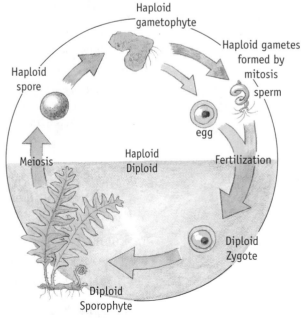

Figure 8-10
Typical Plant Life Cycle.

Prophase I

This is the longest part of meiosis I. As in mitosis, the chromosomes condense, the nucleoli and nuclear membrane disappear, and a spindle apparatus forms.

In prophase I, the homologous chromosomes come together in the process of **synapsis.** They are so close to each other that the chromatids may overlap in a process called **crossing over.** The actual site of crossing over is called a **chiasma.** Two chromosomes may exchange chromatid segments at the chiasma, as shown in Figure 8-11.

Q9. What happens to the genes during crossing over?

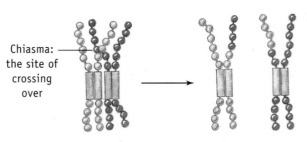

Chiasma: the site of crossing over

Figure 8-11
Synapsis. Homologous chromosomes may crossover, exchanging equivalent pieces of DNA.

The exchange of genetic material during crossing over is another way in which sexually reproducing organisms recombine genes to increase genetic diversity.

Pair your homologous chromosomes and allow a cross over event to occur: overlap a portion of two homologous chromatids, and exchange the beads past the chiasma. Draw chromosomes before and after the event.

Metaphase I

During metaphase I, the chromosomes move to the center of the cell, as in mitosis. Unlike mitosis, the two homologous chromosomes stay next to each other. The centromeres of each chromosome attach to spindle fibers from the opposite poles.

Line up your chromosomes in the center of the cell: one pair of homologous chromosomes next to each other, and the other pair below the first pair. In your drawing, add the spindle fibers, attaching each chromosome from a homologous pair to the opposite pole.

Anaphase I

In anaphase I, the sister chromatids of a chromosome are **not** separated; instead, the whole chromosome moves to the pole. One pair of sister chromatids moves to one pole; the other pair of sister chromatids moves to the other pole.

Arrange your chromosomes into anaphase I and draw them.

Telophase I

Telophase I is similar to telophase in mitosis. The mitotic apparatus disappears, and the nuclear membranes reform around two new nuclei. Cytokinesis occurs, so there are two new cells. The chromosomes may remain partially condensed, and the cell goes straight into meiosis II.

Draw the arrangement of chromosomes in telophase I. In your drawing, include the nuclear membrane and plasma membrane of each new cell.

Q10. You may notice that there is another combination of chromosomes that could have occurred during meiosis I, producing two cells different from the ones you drew. Draw the other possible combination of chromosomes.

Meiosis II

Meiosis II is similar to mitosis except there is no DNA replication before Mitosis II. Using your pop-bead chromosomes, simulate meiosis II in your two cells.

Prophase II

Chromosomes condense and the spindle apparatus forms.

Q11. What happens to the nuclear membrane?

Metaphase II

Chromosomes align along the center of the cell; microtubules connect to the sister chromatids at the kinetochores.

Anaphase II

Sister chromatids separate at the centromere and are pulled toward the poles of the cell.

Telophase II

Nuclear membranes form around the chromosomes; chromosomes begin to appear diffuse again.

Cytokinesis

Plasma membranes divide the cells.

Draw the resulting cells, adding nuclear membranes and plasma membranes. These are gametes.

Q12. How many cells did you begin with? How many do you have now?

Q13. Are these cells identical to the cell you began with? If not, how are they different?

Lab 9   Genetics

This lab accompanies Chapter 9 of *Asking About Life.*

Materials/Equipment

Pop beads, about 70 each of two colors (red and yellow), unattached

Table for results, on blackboard or overhead projector

PTC taste paper

Objectives

1. Understand the relationship between genotype and phenotype

2. Perform simple crosses to see how and in what ratios recessive and dominant genotypes and phenotypes are produced

3. Understand Mendel's principles of segregation and independent assortment

4. Look at the distribution of human dominant and recessive traits in the members of the class

Introduction

Genetics is the study of inheritance. A zygote contains genetic information from each of its parents but becomes a unique individual. All physical and behavioral traits that constitute an individual are its **phenotype.** Your hair color and your height are both phenotypic traits, qualities that you can see. In contrast, a **genotype** refers to the gene or genes that influence a phenotypic trait.

The genotype of an organism is set at fertilization: two gametes unite, combining their DNA. The phenotype, however, is a product of the genotype as well as the environment in which the individual grows up.

After studying meiosis and the production of haploid gametes, you know that a diploid cell has two sets of chromosomes, one from each parent. Although homologous chromosomes are composed of genes for the same traits in the same order, they may have different forms of the same gene. For instance, at the site of the gene that influences the presence of freckles, one chromosome may have a gene that results in freckles, while its homolog may have a gene that results in no freckles. Different versions of the same gene are called **alleles.** If you are **homozygous** (*homo* = same) for an allele, it means that you have two copies of that allele: maybe you have two alleles for no freckles. If you have two different alleles for a single gene (one for freckles, one for no freckles), then you are **heterozygous** (*hetero* = different) for that gene.

If a heterozygote shows the phenotype of only one of its alleles, that allele is **dominant;** and the other is **recessive.** In the example of freckles, if you have one allele for freckles and the other for no freckles, you will have freckles (your phenotype) because the gene that results in freckles is dominant. The gene that results in no freckles is recessive. You will have a phenotype of no freckles only if you are homozygous for no freckles—that is, you have two copies of the gene that results in no freckles. Examine Figure 9-1, where *F* represents the dominant allele for freckles and *f* represents the recessive allele for no freckles.

Some genes exhibit **partial dominance**, in which neither allele shows dominance. Some plants show partial dominance for flower color. For instance, if a snapdragon

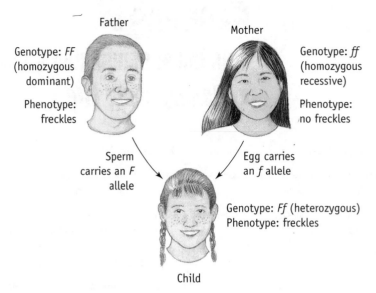

Figure 9-1
Freckles. Genotypes and phenotypes of parents and their child.

homozygous for red flowers is crossed with a snapdragon homozygous for white flowers, the resulting offspring will have pink flowers, a combination of red and white.

In this lab, you will learn how gametes come together in the formation of offspring and how the genes that an individual inherits (its genotype) influence its phenotype.

Mendelian Genetics

When Gregor Mendel began his experiments with pea plants in 1856, genes and chromosomes had not been discovered. However, he discovered the basic principles of genetics through his analysis of phenotypic ratios. He would cross individuals of certain phenotypes and use the phenotypes of the resulting offspring to determine the genetic makeup of the parents and the process by which their characteristics were passed on.

The simplest tests Mendel performed involved the inheritance of single traits that had only two forms. We now know that he was examining traits controlled by single genes that had only two alleles.

You will perform several crosses using colored pop beads to represent two different alleles of a gene. Red will represent the dominant allele, designated *R*; yellow will be the recessive allele, *r*. Work in pairs: one lab partner will be homozygous for the dominant allele (with two red beads for the *RR* genotype); the other will be homozygous for the recessive allele (with two yellow beads for the *rr* genotype). The two beads represent the alleles on homologous chromosomes. When gametes are formed by meiosis, each gamete receives one chromosome of a homologous pair, so a single bead may represent a gamete and allele on that chromosome. You and your partner will "mate" by exchanging gametes.

F1 Generation

This cross may be written as "*RR* × *rr*." You and your partner are the **parental**, or **P, generation**, and your progeny will be the **first filial**, or **F1, generation**.

1. Take your two gametes (beads), shake them in your cupped hands, and choose one without looking. Your partner should do the same. On the data sheet, write the two alleles chosen: this is the genotype of the first offspring.

2. Take back your gamete so that you have the same genotype with which you started, and perform this mating, or **cross**, nine more times. Record the genotypes of your progeny on the data sheet. Also record these results on the table on a blackboard or overhead projector so that the results from the whole class can be compared.

    Q1. What are the genotypes produced by this crossing?

    Q2. One parent exhibited the dominant phenotype, and the other exhibited the recessive phenotype. What phenotypes are exhibited by the F1 generation?

    Q3. Is it possible to produce offspring with the recessive phenotype from this cross?

F2 Generation

Members of the **F2 generation** are the progeny of the F1 generation. You can predict the genotypes that will occur in the F2 generation by making a **Punnett square:** Write the possible gametes from one parent along the top of the square and the possible gametes from the other parent along the left side of the square. Inside the squares, write the genotypes produced by each combination of gametes. A Punnett square for this mating has been started for you: finish by completing the empty boxes.

Rr × Rr

	R	r
R	RR	
r		

The results of your Punnett square should predict that the offspring will occur in a **genotypic ratio** of 1:2:1 (1 homozygous-dominant:2 heterozygous:1 homozygous-recessive).

1. Change beads so that you and your partner each have the F1 genotype.

2. "Mate" with your partner 10 times—make sure you choose gametes without looking. Record the genotypes of the progeny on the data sheet and at the front of the classroom.

    Q4. Looking at the data from the whole class, is the genotypic ratio 1:2:1, as predicted by the Punnett square?

    Q5. Why is it better to use combined data from the whole class (instead of individual results from one group) to figure this ratio?

    *Q6. What is the **phenotypic ratio**?*

    Q7. What happened to the recessive phenotype?

Testcrossing

When Mendel performed crosses, he could only look at the phenotypes of the offspring. From these he had to deduct what was happening on a genetic level. When he crossed an F1 generation like you just did, he found that in a quarter of the offspring, the recessive phenotype reappeared. He reasoned that each sexually reproducing organism has two genes for each characteristic that **segregate** (separate) during the production of gametes. This was his **principle of segregation.** Although the F1 organisms did not exhibit the recessive trait, they must carry the recessive allele, and half of the gametes produced by F1 individuals would carry that recessive allele. To test this theory, he performed a **testcross:** he crossed an individual from the F1 generation with an individual from the parental stock (P) that showed the recessive phenotype.

> Q8. *What is the genotype of the F1 individual? Of the recessive P individual?*

> Q9. *Predict the results of such a testcross. You may want to draw a Punnett square. Write the predicted genotypic and phenotypic ratios on the data sheet.*

 Remember that Mendel could only look at phenotypes.

> Q10. *What did the results of this testcross tell Mendel about his theory that genes segregate during the production of gametes?*

Any time you have an individual showing a dominant phenotype, you may perform a testcross to reveal its genotype.

1. Change beads again so that one lab partner has a homozygous recessive genotype *(rr)*; the other lab partner should get beads for a dominant phenotype (either *RR* or *Rr,* but do not let the lab partner with the recessive genotype see which combination you have).

2. Perform 10 crosses and record the resulting genotypes on the data sheet.

> Q11. *What is the genotype of the "unknown" parent, based on the results of this testcross?*

> Q12. *Why is a homozygous recessive individual used in a testcross instead of a homozygous dominant individual?*

Two-Factor Cross

Another type of experiment Mendel performed on pea plants tested the inheritance of two traits. This is called a two-factor cross. In one test, he crossed a pea plant that produced round, yellow peas *(RRYY)* with a plant that produced wrinkled, green peas *(rryy)*. Mendel already knew that yellow was dominant over green and that round peas were dominant over wrinkled peas.

As expected, the F1 generation produced all round, yellow peas *(RrYy)*. But what happened in the F2 generation? When Mendel allowed the F1 plants to self-pollinate, the F2 generation had a phenotypic ratio of 9:3:3:1 (9 round-yellow: 3 wrinkled-yellow:3 round-green:1 wrinkled-green). Mendel could explain these results if each F1 plant produced four types of gametes: *RY, Ry, rY,* and *ry*, as shown in Figure 9-2.

From this information, Mendel established the **principle of independent assortment:** each pair of genes is distributed independently in the formation of gametes. The gametes produced by the parental generation were all either *RY* or *ry*, but the *R* and *Y* and the *r* and *y* in the F1 generation were distributed into gametes independently. The *R* allele was just as likely to be combined with the *Y* allele as it was to be combined with the *y* allele.

TWO-FACTOR CROSS

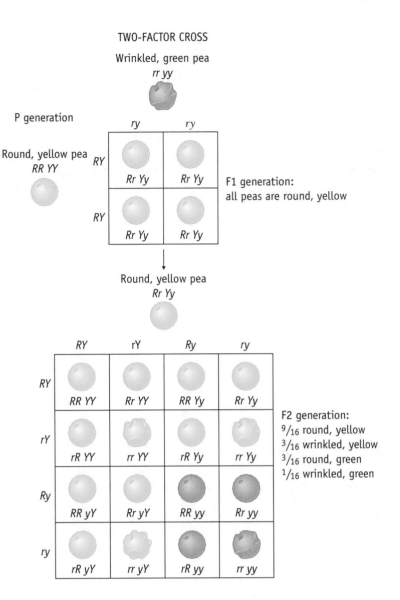

Wrinkled, green pea
rr yy

P generation

Round, yellow pea
RR YY

	ry	*ry*
RY	*Rr Yy*	*Rr Yy*
RY	*Rr Yy*	*Rr Yy*

F1 generation:
all peas are round, yellow

Round, yellow pea
Rr Yy

	RY	*rY*	*Ry*	*ry*
RY	*RR YY*	*Rr YY*	*RR Yy*	*Rr Yy*
rY	*rR YY*	*rr YY*	*rR Yy*	*rr Yy*
Ry	*RR yY*	*Rr yY*	*RR yy*	*Rr yy*
ry	*rR yY*	*rr yY*	*rR yy*	*rr yy*

F2 generation:
$^9/_{16}$ round, yellow
$^3/_{16}$ wrinkled, yellow
$^3/_{16}$ round, green
$^1/_{16}$ wrinkled, green

Figure 9-2
Two-Factor Cross Tests Two Traits at Once. Mendel crossed round, yellow peas with wrinkled, green peas. In the first (F1) generation, all the peas were round, yellow heterozygotes. When he crossed members of the F1 generation, he got an assortment of peas.

Work through a two-factor cross using tomatoes.

1. In tomatoes, red fruit color (*R*) is dominant over yellow fruit color (*r*) and oblong fruit shape (*L*) is dominant over normal fruit shape (*l*), as shown in Figure 9-3.

2. On the data sheet, draw a Punnett square, showing a cross between a tomato plant homozygous for red, oblong fruit and a tomato plant homozygous for yellow, normally shaped fruit (see Figure 9-4).

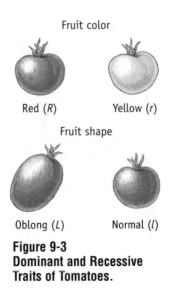

Fruit color

Red (*R*)   Yellow (*r*)

Fruit shape

Oblong (*L*)   Normal (*l*)

Figure 9-3
Dominant and Recessive Traits of Tomatoes.

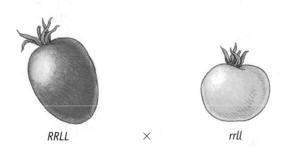

RRLL   ×   *rrll*

Figure 9-4
Parental Cross Involving Two Factors. Fruit shape and fruit color.

Q13. What types of gametes can the plant with red, oblong fruit produce?

Q14. What types of gametes can the plant with yellow, normal fruit produce?

Q15. What are the genotypes of the F1 generation produced from this cross?

Q16. What are the phenotypes of the F1 generation produced from this cross?

3. An F1 individual is self-pollinated.

    Q17. What types of gametes can this individual produce?

4. Fill in the Punnett square on the data sheet for this cross. First write the types of gametes along the top and left side of the square, then combine the gametes to produce the F2 generation in the square.

    Q18. How many of the F2 generation are red and oblong?

    Yellow and oblong?

    Red and normal?

    Yellow and normal?

    Q19. Does the F2 generation show the expected phenotypic ratio of 9:3:3:1?

    Q20. How do these results support Mendel's principle of independent assortment?

Human Genetic Traits

When Mendel crossed his pea plants, he was careful to choose traits that were controlled by just two alleles and were clearly dominant or recessive. These characteristics made it possible for Mendel to demonstrate the basics of genetic inheritance but not more complex situations. Most phenotypic traits are controlled by many different genes. (This is discussed in depth in Chapter 14 of the text.) However, several human traits exhibit the dominant and recessive behavior demonstrated by the crosses with the beads. Some of these traits are illustrated in Figure 9-5.

If you are recessive for a characteristic, you automatically know your genotype. If you exhibit the dominant phenotype, you may be able to determine your genotype if you know the phenotypes of your parents, siblings, or children. Circle your genotype on the table on the data sheet, if you can. Record your phenotype and the phenotypes of the rest of the students in the class on the data sheet. You or your instructor can then determine the percentage of students in the class with each trait.

A. **Free earlobes:** Unattached earlobes are dominant over attached earlobes.

B. **Widow's peak:** A distinctive downward point to the hairline on the forehead is dominant over a straight hairline.

C. **Tongue rolling:** Rolling the sides of the tongue upward is not a learned trait—those who cannot do it are recessive for this characteristic.

D. **Freckles:** These are a dominant characteristic; lack of freckles is a recessive characteristic.

E. **Mid-digital hair:** The presence of hair on the middle joint of your fingers is a dominant trait, even if it's only on one finger.

F. **Facial dimples:** Dimples at the corner of the mouth are a dominant trait.

G. **Finger interlacing:** Fold your hands together with your fingers interlaced. If your left thumb lies over your right thumb, you are dominant for this trait; if your right thumb lies over the left, you are recessive for this trait. One position will feel more natural for you.

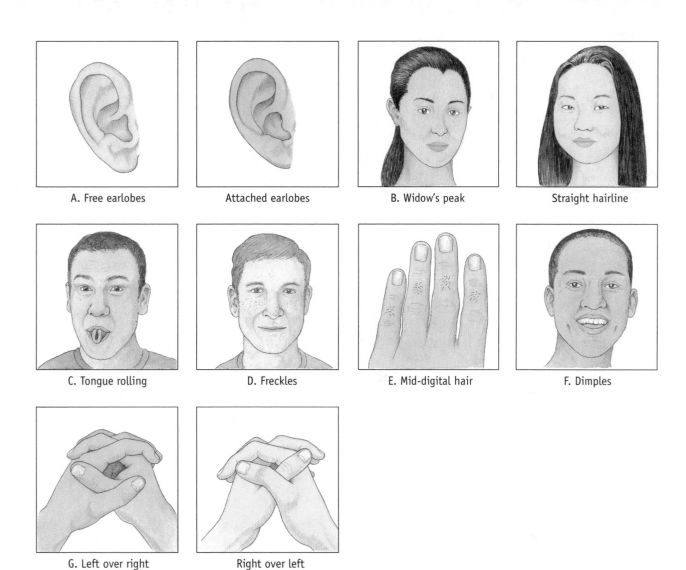

Figure 9-5
Human Genetic Traits.

H. **PTC tasting:** Some people can taste the chemical phenylthiocarbamide (PTC); others cannot. If the PTC paper tastes really bitter, you are dominant for this trait; if it just tastes like paper, you are recessive for this trait.

Q21. *Do the dominant phenotypes occur more frequently in your class?*

Q22. *If you can roll your tongue but your brother and mother cannot, what must your father's genotype be? Can he roll his tongue?*

Q23. *You are trying to establish paternity in a court case. If the mother has attached earlobes and freckles, does the freckled child belong to the man with no freckles and free earlobes or to the man with freckles and attached earlobes? Are the child's earlobes attached or free? Explain your answers, keeping the genotypes in mind.*

Lab 10   DNA
This lab accompanies Chapters 10 and 13 of *Asking About Life.*

6?4

Materials/Equipment

Three-dimensional DNA model

Objectives

1. Develop an understanding of the structure and properties of DNA, based on observations and manipulations

2. Observe and participate in agarose gel electrophoresis; understand some applications of such a technique

3. Understand some implications of DNA technology

4. Extract DNA from an onion to understand that all cells contain DNA

Introduction

In the lab, you have seen the site of DNA (the nuclei of eukaryotic cells) and the condensed chromosomes of a cell as it undergoes mitosis. You have manipulated beads that represented genes in chromosomes to simulate mitosis and meiosis. And you have studied **genetics:** how genes are transmitted from one generation to the next. But you have yet to actually look at DNA, the hereditary material contained in almost every living cell, from bacterium to broccoli to elephant. Part of the problem is that DNA is very small—only 2 nm in diameter. A human hair, in contrast, is about 50 μm wide—that is, 25,000 times the width of a DNA molecule. The 2 m (about 6 feet!) of DNA in each of your cells is tightly wound with histone proteins into chromosomes, as shown in Figure 10-1.

**Figure 10-1
How Does DNA Pack Tightly into the Nucleus?**
Histone proteins pack DNA into small, tight nucleosomes; strings of nucleosomes fold and loop to form chromosomes small enough to fit into the nucleus.

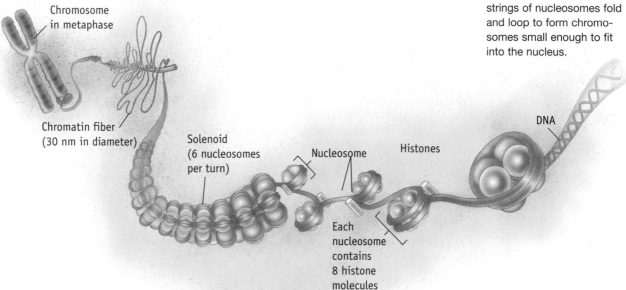

Chromosome in metaphase

Chromatin fiber (30 nm in diameter)

Solenoid (6 nucleosomes per turn)

Nucleosome

Histones

DNA

Each nucleosome contains 8 histone molecules

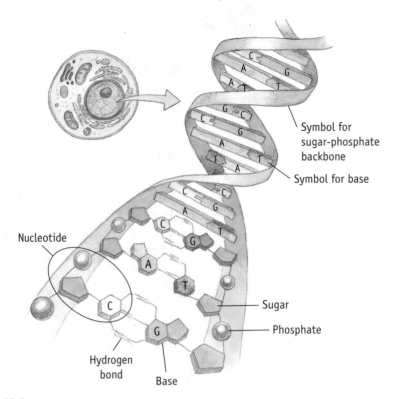

Figure 10-2
In a Eukaryotic Cell, DNA Is Found in the Nucleus. The double helix structure of DNA consists of two sugar-phosphate backbones with attached nucleotide bases. The nucleotide bases on opposite strands are linked by hydrogen bonds.

The structure of DNA was not discovered until 1953—precisely because DNA cannot be viewed directly. The methods and story behind this discovery are in the introduction to Chapter 10 of the text. We now know that DNA is the shape of a **double helix,** two parallel strands of DNA, linked by hydrogen bonds and twisted. DNA stands for **deoxyribonucleic acid.** It is a polymer composed of **nucleotides,** one of the four types of molecular building blocks. *Do you remember, from Lab 2, what the other three building blocks are?*

Each nucleotide is composed of phosphate + sugar (deoxyribose) + nitrogen base (adenine [A], guanine [G], cytosine [C], or thymine [T]). The sugar and phosphate of each nucleotide make up the **sugar-phosphate backbone** of the two DNA strands; the bases connect the two backbones by hydrogen bonding. Each A bonds with a T on the opposite strand, and each G pairs with a C on the opposite strand, as shown in Figure 10-2.

One of the most compelling pieces of evidence for evolution is the universality of the genetic code: bacteria, ferns, lizards, and humans all have DNA that codes for proteins similar in structure and function. Consequently, knowledge gained from the study of the genomes of organisms such as yeasts and bacteria can be used to learn more about the human genome.

In this lab, you will study the structure of DNA, learn some basic molecular techniques, and observe DNA that you have isolated from an onion.

DNA Structure

Examine the DNA model, if available. Note its three-dimensional double helix structure.

1. Locate the sugar-phosphate backbone on each strand.
2. Identify the A–T bonds and the G–C bonds.

Q1. How many hydrogen bonds constitute each of these bonds?

3. Examine one of the strands of the double helix. You know that genes are portions of DNA that carry the information necessary for the production of proteins. Each sequence of three nucleotides on a strand of DNA codes for an amino acid; the sequence of amino acids determines which polypeptide is produced by a gene.

Agarose Gel Electrophoresis

Scientists employ several methods to determine the molecular composition of chemical mixtures. Macromolecules may be separated and categorized according to their size, shape, and electrical charge. **Electrophoresis** uses an electrical current to separate molecules with different charges: The mixture is placed in a conductive environment between the **cathode** (negative charge) and the **anode** (positive charge). The molecules with a positive charge will be drawn toward the cathode, and the molecules with a negative charge will migrate toward the anode.

Agarose is a carbohydrate derived from seaweed and can be used to separate molecules by size and shape. Made into a gel, the agarose has a texture similar to gelatin. At a molecular level, the agarose molecules form a three-dimensional network through which other molecules may pass. Imagine you and a cat are starting at one end of a large room packed with furniture, trying to get to the food at the other end of the room. The cat can run under, between, and over the furniture; because you are larger, you have fewer options. You must find your way around and between the big pieces of furniture. Like the furniture, the agarose provides an obstacle course for molecules. The small molecules can make it through the gel more rapidly than the larger molecules.

Agarose gel electrophoresis combines these two techniques to pull molecules through an agarose gel using an electric current, thereby separating them by size, shape, and electric charge, as shown in Figure 10-3.

Your instructor will demonstrate how to pour an agarose gel, load the gel with DNA, and run the gel through an electrophoresis chamber. If time and materials allow, you may be able to participate in these steps. You will be asked to analyze the results on the gel.

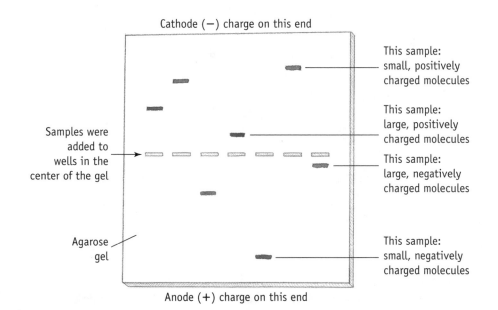

Cathode (−) charge on this end

Samples were added to wells in the center of the gel

Agarose gel

Anode (+) charge on this end

This sample: small, positively charged molecules

This sample: large, positively charged molecules

This sample: large, negatively charged molecules

This sample: small, negatively charged molecules

Figure 10-3 Example of Agarose Gel Electrophoresis. Seven different samples were placed in wells in the center of the gel. Each sample migrated through the gel depending on the charge and size of the molecules.

Restriction Enzymes

Preparation of the DNA was performed before the lab. Bacterial DNA was purchased from a biological supply company. DNA molecules are large; if they were applied to an agarose gel in a whole form, they would move slowly through the gel and appear as one band on the gel after electrophoresis. One important tool used in DNA technology is a group of bacterial enzymes called **restriction enzymes.** Restriction enzymes recognize specific sequences of nucleotides and cut the DNA strand at these locations, as shown in Figure10-4 (A). In nature, these enzymes protect the bacterial cell from being infected by viruses and other mobile genes by recognizing and cutting up the foreign DNA.

Because a molecule of DNA contains thousands to millions of base pairs, the sequence of nucleotides cut by a particular restriction enzyme is likely to occur in several locations in that molecule. The DNA will be cut at each of those sites. The resulting fragments will separate during electrophoresis, based on their sizes, and will appear as a series of bands on the gel.

Restriction enzymes are named for the bacterium from which they were first isolated. For instance, the *Eco*RI shown in Figure 10-4 (B) was isolated from *E. coli.* The "R" refers to the strain of *E. coli,* and the "I" refers to the order in which this enzyme was discovered.

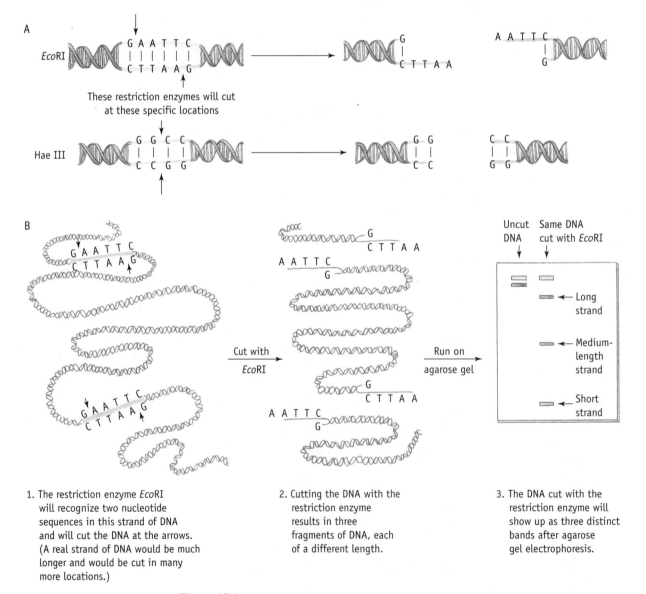

1. The restriction enzyme *Eco*RI will recognize two nucleotide sequences in this strand of DNA and will cut the DNA at the arrows. (A real strand of DNA would be much longer and would be cut in many more locations.)

2. Cutting the DNA with the restriction enzyme results in three fragments of DNA, each of a different length.

3. The DNA cut with the restriction enzyme will show up as three distinct bands after agarose gel electrophoresis.

Figure 10-4
A. How restriction enzymes work. **B.** Cutting a strand of DNA with *Eco*RI, a restriction enzyme.

In this demonstration, three or four different restriction enzymes were used to cut up (or **digest**) the same type of bacterial DNA. Because each restriction enzyme cuts the DNA strand at a different site, digestion by one restriction enzyme will result in a pattern of short and long fragments different from the pattern that results from digestion by another restriction enzyme. Your instructor will tell you which restriction enzymes were used in this demonstration.

Demonstration

1. Preparing the gel: Molten agarose is poured into a gel mold with a comb (see Figure 10-5). The gel will need to harden for about 15 minutes before it is ready to use.

2. Loading the wells: When the gel is ready, the comb is removed, leaving many wells into which the samples can be loaded.

    a. The gel is placed into the electrophoresis chamber, and an electrically conductive buffer solution is added to the chamber.

    b. A loading dye has been added to the DNA samples for two reasons: the samples are visible, and the dye particles are heavy, allowing the samples to sink to the bottom of the wells instead of float away into the buffer. The samples are loaded into the gel. It is important to keep track of which sample was loaded into which well.

    Q2. With what will the control well be loaded?

3. Running the gel: Because DNA molecules carry a negative charge, the samples will migrate toward the anode (the positive charge). A power supply attached to the electrophoresis chamber will be turned on, and the gel will run for about 45 minutes, until the loading dye has moved about 2/3 of the way down the gel. **The power supply carries a high voltage. Do not remove the lid of the chamber when the power supply is on.**

> ⚠️ **While you wait for the gel to run, begin the onion DNA experiment.**

Comb

Gel mold

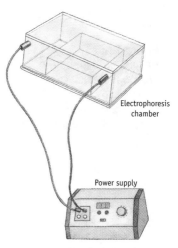
Electrophoresis chamber

Power supply

Figure 10-5
Electrophoresis Equipment.

4. Staining the samples: The DNA samples are not visible unless they are stained. The instructor (or a student group) will place the gel in a staining tray and add a blue stain that specifically stains DNA, allowing you to see the fragments. The gel must be stained for about 15 minutes and then destained in deionized water for 15–20 minutes, with a water change during that time. The DNA fragments will become visible.

5. Analyzing the samples: Examine the banding patterns on the stained gel.

    Q3. What does the control sample look like? How far did it travel? Why?

    Q4. Can you tell that different restriction enzymes were used for the other samples?

    Q5. What causes the different banding pattern of each sample?

    Q6. What would have happened if the electrodes (positive and negative charges) had been reversed?

DNA Fingerprinting

In the agarose gel electrophoresis demonstration, different restriction enzymes were used to cut the same DNA. In many electrophoresis applications, restriction enzymes are used to cut DNA from different individuals. Every individual (except

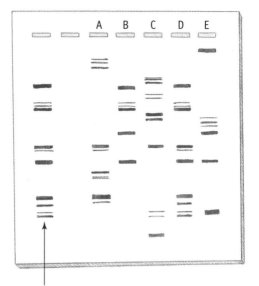

DNA from hair sample found at crime scene

Figure 10-6
Results of Agarose Gel Electrophoresis. Comparing a hair sample from a hypothetical crime scene to hair samples takes from suspects A, B, C, D, and E.

identical twins) has a slightly different nucleotide sequence, so a restriction enzyme will produce a different banding pattern in every person. The difference in banding patterns is called **restriction fragment length polymorphism,** or RFLP (*poly* = many, *morph* = form). Your specific banding pattern is your **DNA fingerprint.** Scientists can use this kind of data to help identify genetic diseases and cancers; DNA fingerprints are also useful in legal cases, such as cases of paternity, murder, or rape. Blood, hair, semen, skin, or other tissues found on a crime scene may be collected and subjected to DNA analysis. The resulting DNA fingerprints may then be compared with the DNA fingerprints of the suspects. This method can help identify criminals with a higher degree of certainty than many other types of evidence. You will learn more about this process in Chapter 14 of the text.

Figure 10-6 shows the results of agarose gel electrophoresis of the DNA from several hair samples related to a hypothetical murder scene. Use this RFLP data to identify which suspect committed the crime.

Q7. Which suspect committed the murder?

Q8. Circle the band representing the smallest DNA fragment on this gel. Explain why it is located where it is on the gel.

Q9. The human genome contains about 3 billion nucleotide pairs, so only highly variable portions of the genome are used for RFLP analysis. Why does it make sense to use highly variable portions of the genome? Conversely, how could it be argued that a conviction based on this type of evidence is not entirely valid?

Q10. What are three processes responsible for the variety in the human genome that makes it possible to have unique DNA fingerprints? (Think back to cell division.)

Onion DNA Extraction

All living cells contain DNA. You have seen chromosomes in animal and plant cells undergoing mitosis. In this exercise, you will start with a whole onion and isolate its DNA. Within an hour, you will have a test tube containing billions of genes.

The steps in Part A prepare the onion for DNA extraction. One onion will provide enough material for the whole class. To get the DNA out of the cells, the cell walls, plasma membranes, and nuclear membranes must be broken down. Next, the DNA will be separated from some of the proteins bound to the DNA in the chromosomes. **These first steps will be divided among the student groups.**

For Part B, each group will take some of the prepared onion and add cold ethanol (an alcohol), which will cause the DNA to come out of solution.

Materials/Equipment

One yellow onion

Two 250 mL beakers

Glass stir rod

Funnel

Cheesecloth

Knife

Cutting board

Balance

100 mL homogenizing medium

5.0 g sodium dodecyl sulfate (SDS)

95% ethanol, ice cold, three or four 50 mL flasks with pipettes

60°C water bath

Thermometer

Ice bath

Blender

3 mL graduated pipette for dispensing onion homogenate

Wash bin for used spooling pipettes and test tubes

Waste beaker

Per group of two or three students:
   One test tube (15×95 mm)
   Spooling pipette

Part A: Onion Preparation

Each step should be performed by a student group.

1. Peel and dice one onion.
2. Weigh 50 g of diced onion and place it in the jar of a blender.
3. Add 100 mL of homogenizing medium to the onion in the blender and put the lid on. Process the onion on medium high for about one minute. The homogenizing medium contains salts that help maintain the structure of the DNA during the isolation process.

   Q11. What does the blender do to help get the DNA out of the cells?

4. Pour the processed onion mixture into a 250 mL flask. Add 5.0 g of SDS and mix well with a glass stir rod. SDS is a detergent that helps dissolve cell membranes and denature proteins.
5. Heat the flask in a 60°C water bath for 12–15 minutes; remove promptly and place the beaker into an ice bath. The heat softens the onion tissues, allowing the SDS and homogenizing medium to penetrate.

   Q12. There are several enzymes present in the nucleus that could interfere with the DNA isolation process. What does the heat treatment do to prevent this interference?

6. Place a thermometer into the flask and let the mixture cool in the ice bath until it reaches 15–20°C (about five minutes). When checking the temperature of the mixture, raise the thermometer slightly until it is suspended in the mixture and not touching the bottom of the flask. Cooling prevents **denaturation** of the DNA, in which the hydrogen bonds holding the two strands together are broken.

   Q13. Why would your temperature reading be inaccurate if you did not raise the thermometer from the bottom of the flask?

7. Filter the mixture using a funnel and four layers of cheesecloth into a clean 250 mL flask, keeping the flask on ice if possible. It may take several minutes for the mixture to go through the cheesecloth.

Part B: Spooling the DNA

To be performed by each group.

1. Transfer 4 mL of the onion mixture to a clean test tube. Gently swirl your spooling pipette in the mixture to get an idea of its texture. Note the color as well. Rinse and dry the spooling pipette before proceeding to the next step.

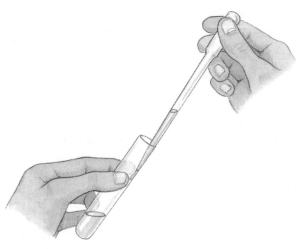

Figure 10-7
Method for Adding Ethanol to the Onion Mixture. Tilt the test tube slightly and add the ethanol down the side of the test tube. Do not agitate the test tube.

2. Slowly add about 2 mL of ice-cold 95% ethanol down the side of the test tube, as illustrated in Figure 10-7.

    The ethanol will form a distinct, clear layer over the yellowish onion mixture. As you add the ethanol, you will notice a new layer forming between the ethanol and the onion mixture. It should be clear and slightly jellylike, with tiny whitish strands. This layer is the onion's DNA.

    DNA is not soluble in the presence of salt and alcohol. Before you added the ethanol, the DNA was in solution with the salts from the homogenizing medium. Adding the ethanol caused the DNA to come out of solution as a solid.

3. Gently swirl the end of the spooling pipette in the DNA layer so that the DNA wraps around the pipette, much like spaghetti being wound around a fork.

    Q14. Describe the appearance and texture of the DNA before and after adding the ethanol.

    Q15. What property of DNA allows it to be spooled around your pipette?

    Q16. The DNA you examined in cells undergoing mitosis looked very different from the DNA in your tube. What are some reasons for this difference? Hint: Think about how this DNA was treated, as well as the form chromosomes take during mitosis.

    Q17. How would the results of this procedure be affected if the SDS was not added? Explain your answer.

    Q18. Why would the DNA in this form run poorly on a gel?

    Q19. What would need to be done to the DNA to prepare it for agarose gel electrophoresis?

4. When you are through with the DNA, empty it into the waste beaker and put the empty test tube and spooling pipette into the wash tub.

Biotechnology

Humans have been altering the genetics of agriculturally useful plants and animals for hundreds of years by selective breeding. New technology is allowing genetic manipulation to occur more rapidly and more precisely.

Biotechnology is the study of techniques that use living things to solve problems and make products. Box 10-1 in the text describes how scientists can determine the sequence of DNA, nucleotide by nucleotide. Until 2000, only the relatively small genomes of several organisms had been **mapped.** But in June 2000, a draft of the entire human genome was completed. Scientists all over the world participated in the Human Genome Project, working to identify all 3 billion nucleotide pairs in the human genome. Besides gaining an understanding of genome organization, control of gene expression, and evolutionary biology, this study has great implications. Identification and mapping of the genes responsible for genetic diseases and cancer may help diagnose, treat, and possibly prevent these conditions. Once we understand the function of each human gene, we can begin to understand human evolution and the common biology we share with all other life.

Another use of DNA technology involves the manipulation, or **engineering**, of plant, animal, and microbial DNA to give those organisms more desirable characteristics. Researchers can use restriction enzymes to cut bacterial DNA and then insert genes into the bacterial genome, causing the bacteria to produce particular enzymes or hormones. Bovine growth hormone, which is produced in this manner, causes increased milk production in cows.

Genetic research is being conducted to develop bacteria that can degrade harmful compounds from oil spills and other toxic waste. Genes can be inserted into the genomes of plants, giving those plants resistance to herbicides, diseases, or pests.

The long-term consequences of manipulating genes are unknown and raise many ethical, moral, and environmental issues. The introduction of foreign genes into wild populations from the pollen of genetically manipulated crop plants is one concern; any kind of genetic manipulation of the human genome is another controversial topic.

> Q20. *The motivation behind a company's investment in biotechnology is the potential for a large financial gain. What is a positive aspect of this motivation? A negative one?*

One genetically engineered food product that made it onto the market was a tomato that had genes from other organisms inserted into its genome to prevent it from ripening too soon. It resulted in an easily harvested, flavorful tomato. These tomatoes did not do well on the market.

> Q21. *Would you have purchased these tomatoes? Explain the reasoning behind your answer.*

Lab 11   Microevolution: How Does a Population Evolve?
This lab accompanies Chapter 16 of *Asking About Life.*

Objectives

1. Grasp the basic mechanics of evolution within a population

2. Calculate allele frequencies and use the Hardy-Weinberg Equilibrium to determine if evolution has occurred

3. Understand how evolution works on genetic variation

4. Understand effects that natural selection can have on genotype and phenotype frequencies

5. Understand how the environment and interactions between species can influence evolution

Introduction

When Charles Darwin published *On the Origin of Species* in 1859, he was not the first one to discover or describe **evolution,** the process by which species change over time. Biologists and philosophers throughout history had recognized the relatedness between organisms. However, none had devised a sound theory for a mechanism by which evolution works. Darwin's revolutionary contribution to the understanding of evolution was his development of the theory of **natural selection** as the mechanism for evolution.

A biological **population** is a breeding group of individuals of one species that inhabits the same area. Natural selection assumes that all organisms produce more offspring than can survive in a world with limited resources. Because each individual in a sexually reproducing species is slightly different, some offspring will be more suited to their environment than others. The offspring that escape death by predation, disease, or starvation long enough to reproduce will pass their characteristics to their offspring. The heritable traits that best suit an individual for its environment will be passed on with a higher frequency than nonadaptive traits, so the advantageous traits will become more common in the population, changing the population over time.

> Q1. Natural selection assumes genetic variation among members of a population. What are three sources of genetic variation in organisms?

In this lab, you will look at some ways that natural selection causes populations of organisms to evolve. You may want to review basic genetics in Chapter 9.

Allele Frequency and the Hardy-Weinberg Equilibrium

Allele frequency indicates how often an allele appears in a population. It is calculated by dividing the number of times the allele is present in the population by the total number of alleles in that population.

$$\text{Allele frequency} = \frac{\text{Number of times an allele is present in a population}}{\text{Total number of alleles in that population}}$$

For example, in a population of 100 fruit flies that carry an eye color gene that has two alleles, *A* and *a*, the dominant allele (*A*) is for red eyes, and the recessive allele (*a*) is for purple eyes.

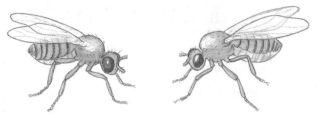

Figure 11-1
Red- and Purple-Eyed Fruit Flies.

*Q2. Twenty-five flies have the genotype **AA**. What is the phenotype of these flies?*

*Q3. Fifty flies have the genotype **Aa**. What is the phenotype of these flies?*

*Q4. Twenty-five flies have the genotype **aa**. What is the phenotype of these flies?*

There are 100 flies, each with two alleles for the eye color gene; therefore, $100 \times 2 = 200$ total alleles in the population.

There are 25 flies with the genotype *AA*. Because each fly has two *A* alleles, there are 50 (2×25) *A* alleles; there are 50 flies with the genotype *Aa*, so these flies have an additional 50 *A* alleles. The total number of times the *A* allele appears in this population is therefore 100 ($50 + 50$). You can calculate its allele frequency as follows:

$$\text{Allele frequency of } A = \frac{\text{Number of times } A \text{ allele is present in population}}{\text{Total number of alleles in population}}$$

$$\text{Allele frequency of } A = \frac{100}{200} = 0.5 \text{ (or 50 percent)}$$

The allele frequency of the *a* allele is also 0.5 (50 percent). Calculate it in the following spaces:

Number of times *a* allele is present in population = _____

$$\text{Allele frequency of } a = \frac{\text{Number of times } a \text{ allele is present in population}}{\text{Total number of alleles in population}}$$

$$= \text{_____} = \text{_____} \%$$

Notice that the sum of the two allele frequencies is $0.5 + 0.5 = 1$ (or 100%), because you have accounted for all, or 100%, of the eye color alleles in the population.

The **Hardy-Weinberg Equilibrium** states that the allele frequencies in a large population will not change unless an evolutionary factor is working, such as:

- Nonrandom mating
- Small population size
- Mutation
- Gene flow (breeding between populations)
- Natural selection

See "What Causes Allele Frequencies to Change?" in the text for descriptions of each of these factors.

The Hardy-Weinberg Equilibrium is useful because it is a baseline against which the evolution of a population can be measured: if the allele frequencies are changing, then one of the listed factors must be at work, and evolution is occurring.

Natural Selection and Allele Frequency

Although any evolutionary force will lead to a change in allele frequencies, natural selection leads to **adaptive changes** in allele frequencies—that is, to changes that increase an individual's chances of surviving and reproducing. Other evolutionary factors cause random changes in allele frequencies: some changes will improve survival and reproduction rates, some will decrease it, and some will not change the overall chances of survival and reproduction.

In this exercise, you will determine the change in allele frequencies on a population (your class) that has a lethal recessive genotype: the genotype *aa* results in a fatal condition that causes death during childhood. The genotypes *AA* and *Aa* both result in healthy individuals.

Materials/Equipment

Pop beads, about 70 each of two colors (yellow and red)

Table on blackboard or overhead projector for students to record mating data

1. Yellow pop beads represent the *A* allele; red beads represent the *a* allele. Each person will start as heterozygous. Record your genotype on the data sheet.

2. Walk around the class and find another person in the class **at random**: tall or short, male or female. Shake your alleles in your cupped hands and choose one without looking while your partner does the same. The resulting two alleles represent the genotype of your first offspring. Record that genotype: _____.

3. Change beads so that you both are heterozygous, and "mate" again. Record that genotype: _____. The two offspring you have just produced will replace you and your partner in the next generation. Each of you should choose one of the offspring, which you will become, and record this genotype on the data sheet. If either offspring has the genotype *aa*, it dies in childhood and cannot reproduce again. So if you have the genotype *aa*, you cannot reproduce again.

4. Now, assume the genotype of your offspring. If you had the genotype *AA* or *Aa*, replace your beads with the appropriate colors.

5. At random, find a new partner. "Mate" twice, recording your new genotype on the data sheet. If you produce an *aa* offspring, you cannot mate in the next generation.

6. Continue to randomly mate four more times until six generations have been completed (unless you produce an *aa* offspring).

7. Report your results to the class and record the data of the class on the data sheet. Calculate the allele frequencies for each generation of the population.

    Q5. Did evolution occur in this population?

    Q6. What happened to the allele frequencies from generation to generation?

    Q7. What happened to the genotype frequencies from generation to generation?

    Q8. Why was it important that you mated randomly?

    *Q9. Does the **a** allele still exist in the population despite selection against it? Why?*

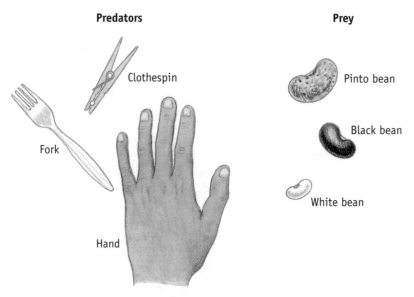

Figure 11-2
Types of Predators and Prey in the Experiment.

Natural Selection in Two Populations

In the previous exercise, you saw how natural selection affected the allele frequency within a population. This exercise illustrates how interactions between different populations can influence the evolution of each population.

Each student will be a member of one of three populations of predators. Each population has a different mechanism by which it captures prey: one species uses one hand, one species hunts with forks, and one species hunts with clothespins. This hunting mechanism is a heritable trait that can be passed on to offspring. The predator's "mouth" in all species is a paper cup held in the other hand.

The prey for these predators consists of three populations of species with different phenotypes: pinto beans, black beans, and white beans.

The prey will be spread across an environment in which the predators will forage for about a minute. The most successful predator and prey species will be better represented in the next generation (they will reproduce). As you go through several generations, you will see how the physical characteristics (phenotypes) of the different species confer advantages or disadvantages to survival.

Materials/Equipment

Pinto beans

Black beans

White beans

Clothespins (about 10)

Plastic forks (about 10)

Paper cups (one per student)

Calculator

Space on blackboard or overhead projector for accumulating data

1. The class will be divided into three equal populations, each of which will be a species of predator. Each species should have a "spokesperson" who will report the total number of prey hunted by that species of predator at the end of each

generation. In addition, one student should total the number of pinto beans captured by all predators during each generation, another student should total the black beans, and a third student should total the white beans.

2. Begin hunting when you are given the signal. Pick up the prey with your feeding tool and place the prey in the cup. Do not push prey into the cup. Feel free to pursue any prey, even if another predator is hunting it. You are hungry and need to compete with other predators to survive.

3. When the time is up, stop hunting and count each of the prey species you captured. Report these numbers to your "spokesperson."

   a. Use the numbers of the prey types gathered from the whole class to fill out the table for Generation 1 on the data sheet.

   b. Calculate the number of surviving pinto bean prey by subtracting the total number of pinto beans consumed from the starting number of pinto beans. Each individual will reproduce, so in the next generation, this number is doubled. Repeat this calculation for the other two prey species.

   c. The number of surviving predators will be based on the total kills for that species. Calculate the total percentage of prey items killed by the predators with forks: find the total number of beans caught by the fork predators, and divide this number by the number of beans at the beginning of the foraging session. This percentage is the percentage of predators that will be fork predators in the next generation. The predators that starve to death will be replaced by more successful predators (the more successful predators reproduce).

      Q10. Based on your experience so far, predict what will happen in this experiment after a few generations.

      Q11. Will the numbers of each predator species stay fairly constant, or will they change? How do you think they will change?

      Q12. Are the species of prey equally easy or difficult to capture? What accounts for any differences?

4. Continue through as many generations as time and resources allow, filling in the tables for each generation on the data sheet.

   Q13. At the beginning of this experiment, there were equal numbers of predator species and of prey species. How have they changed?

   Q14. Do the phenotypes of the predators and prey affect their ability to survive? If they do, explain how.

   Q15. How did competition between populations of predators affect each predator's ability to survive? How would the forks and clothespins have fared if the hands had not been present?

   Q16. How would a prey population become extinct in this experiment? A predator population?

   Q17. Did any environmental factors confer an advantage on any of the prey species? If yes, describe these factors.

   Q18. What would happen in this experiment if the environment changed? For example, how do you think the results would change if the predators foraged first on asphalt and then on dry dirt? How is environmental change related to natural selection?

   *Q19. This exercise is a **model** that simulates what happens when two populations of species interact. What are two ways in which this experiment is lacking—in what ways is it unrealistic?*

Microevolution
Data Sheet

Introduction

Q1. *Natural selection assumes genetic variation among members of a population. What are three sources of genetic variation in organisms?* _____

_____

_____

Allele Frequency and the Hardy-Weinberg Equilibrium

Q2. *Twenty-five flies have the genotype* **AA**. *What is the phenotype of these flies?* _____

_____

Q3. *Fifty flies have the genotype* **Aa**. *What is the phenotype of these flies?* _____

_____

Q4. *Twenty-five flies have the genotype* **aa**. *What is the phenotype of these flies?* _____

_____

Natural Selection and Allele Frequency

	YOUR GENOTYPE
Original genotype	
1st Generation	
2nd Generation	
3rd Generation	
4th Generation	
5th Generation	
6th Generation	

	NUMBER OF PEOPLE WITH THESE GENOTYPES			NUMBER OF INDIVIDUAL ALLELES		ALLELE FREQUENCY	
	AA	Aa	aa	A	a	A	a
Original genotype							
1st Generation							
2nd Generation							
3rd Generation							
4th Generation							
5th Generation							
6th Generation							

Q5. *Did evolution occur in this population?* _____

_____

_____

Q6.   What happened to the allele frequencies from generation to generation? _____

_____

_____

Q7.   What happened to the genotype frequencies from generation to generation? _____

_____

_____

Q8.   Why was it important that you mated randomly? _____

_____

_____

Q9.   Does the **a** allele still exist in the population despite selection against it? Why? _____

_____

_____

Natural Selection in Two Populations

Q10. Based on your experience so far, predict what will happen in this experiment after a few generations. _____

_____

_____

Q11. Will the numbers of each predator species stay fairly constant, or will they change? How do you think they will

change? _____

_____

_____

Q12. Are the species of prey equally easy or difficult to capture? What accounts for any differences? _____

_____

_____

Generation 1

		PINTO BEANS	BLACK BEANS	WHITE BEANS		
	STARTING NUMBER OF PREY					
	STARTING NUMBER OF PREDATORS				TOTAL KILLED	% KILLED
Clothespin						
Hand						
Fork						
	Total prey killed					
	Number of surviving prey					

Generation 2

		PINTO BEANS	BLACK BEANS	WHITE BEANS		
	STARTING NUMBER OF PREY					
	STARTING NUMBER OF PREDATORS				TOTAL KILLED	% KILLED
Clothespin						
Hand						
Fork						
	Total prey killed					
	Number of surviving prey					

Generation 3

		Pinto beans	Black beans	White beans	Total killed	% killed
	Starting number of prey					
	Starting number of predators					
Clothespin						
Hand						
Fork						
	Total prey killed					
	Number of surviving prey					

Generation 4

		Pinto beans	Black beans	White beans	Total killed	% killed
	Starting number of prey					
	Starting number of predators					
Clothespin						
Hand						
Fork						
	Total prey killed					
	Number of surviving prey					

Q13. At the beginning of this experiment, there were equal numbers of predator species and of prey species. How have they changed? _____

_____

_____

Q14. Do the phenotypes of the predators and prey affect their ability to survive? If they do, explain how. _____

_____

_____

Q15. How did competition between populations of predators affect each predator's ability to survive? How would the forks and clothespins have fared if the hands had not been present? _____

_____

_____

Q16. How would a prey population become extinct in this experiment? A predator population? _____

_____

_____

Q17. Did any environmental factors confer an advantage on any of the prey species? If yes, describe these factors.

_____

_____

_____

Q18. What would happen in this experiment if the environment changed? For example, how do you think the results would change if the predators foraged first on asphalt and then on dry dirt? How is environmental change related to natural selection? _____

_____

_____

Q19. This exercise is a **model** that simulates what happens when two populations of species interact. What are two ways in which this experiment is lacking—in what ways is it unrealistic? _____

_____

_____

Questions

Q20.   Why must a trait be heritable to be evolutionarily advantageous for a species? _____

_____

_____

Q21.   What is the importance of genetic diversity in natural selection? How do meiosis and sexual reproduction fit in?

_____

_____

_____

Q22.   What are the sources of genetic change in populations of an organism that reproduces primarily by mitosis? You may

consult the list of evolutionary agents under "Allele Frequency and the Hardy-Weinberg Equilibrium" for help.

_____

_____

_____

Q23.   Some experts argue that pesticides and antibiotics should not be used when they are not absolutely necessary. Explain

their reasoning, based on principles from this lab. _____

_____

_____

In 1846 and 1847, a water mold infected and destroyed the potato crops of Ireland, causing the human population to fall from 8.5 million to 6.5 million in three years, owing to death and emigration. One major reason for the potato famine was the lack of genetic variation among potato crops grown throughout Ireland.

Q24.   Explain how lack of genetic variation is dangerous to the survival of a population, even if the population is well

adapted to its environment. _____

_____

_____

Lab 12   Bacteriology and Epidemiology
This lab accompanies Chapter 20 of *Asking About Life.*

Materials/Equipment

Compound microscopes

Prepared slides:

    Bacillus

    Coccus

    Spirillum

Live specimens or prepared slides of various cyanobacteria

Several bacterial spread plates, with disks of potentially antibacterial substances (such as bleach, mouthwash, alcohol, hydrogen peroxide, and iodine)

Food handling gloves (no powder), one per student

Weight boats or small containers, numbered, one per student:

    All but one with several milliliters of deionized water

    One with several milliliters of 30% starch solution

7% Lugol's iodine in dropper bottle, for instructor use

Nutrient agar or tryptone soy agar (TSA) plates, one per student

Sterile swabs

Tubes of sterile saline or deionized water, one per student

Autoclave bag

Wax pencils

37°C incubator

Objectives

1. Learn the different cell shapes of bacteria

2. Understand the importance of bacteria in the ecosystem

3. Understand the benefits and dangers of antibacterials, and see what common products have antibacterial properties

4. See how disease can be spread and how an epidemiologist could track its course

5. Understand the diversity and abundance of microorganisms and how they are grown in the lab

Introduction

Bacteria are **prokaryotes** (*pro* = before, *karyon* = nucleus), in the kingdom **Eubacteria.** Prokaryotic cells are much smaller than eukaryotic cells (0.4–5 μm vs. 10–100 μm). These are the simplest living organisms: they have no membrane-bound nucleus, and their DNA is not associated with proteins to form chromosomes. Instead, their DNA

Figure 12-1
Prokaryotic Cells Have No Membrane-Bound Organelles. This cyanobacterium's DNA is found in a region called the nucleoid, not in a nucleus. However, a prokaryotic cell may have a folded plasma membrane to increase its surface area for respiration, photosynthesis, and DNA replication.

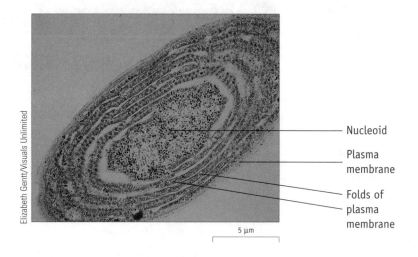

Nucleoid

Plasma membrane

Folds of plasma membrane

Elizabeth Gentt/Visuals Unlimited

5 μm

exists in a circular form and is concentrated in a mass called a **nucleoid,** shown in Figure 12-1. Although prokaryotes may have a folded plasma membrane to increase surface area for respiration, photosynthesis, and/or DNA replication, they completely lack membrane-bound organelles.

Despite their simplicity of organization, prokaryotes—and, in particular, bacteria—are the most numerous, diverse, and ancient organisms. Bacteria live in and on every kind of organism, as well as in oceans, deserts, ice fields, and hot springs. Mitochondria and chloroplasts (one or both of which occur in almost all eukaryotic cells) are similar in size to bacteria and have their own DNA, which is circular and organized into nucleoids. Many biologists believe that mitochondria and chloroplasts originated from free-living prokaryotes that were ingested, but not digested, by predatory cells. An association of two organisms in which one lives inside the other is called **endosymbiosis.** Figure 18-15 in the text shows how this type of association may have begun.

Bacterial Cell Shape

Bacterial cells are commonly one of three shapes: **bacillus** (rod-shaped), **coccus** (spherical), or **spirillum** (spiral), as shown in Figure 12-2.

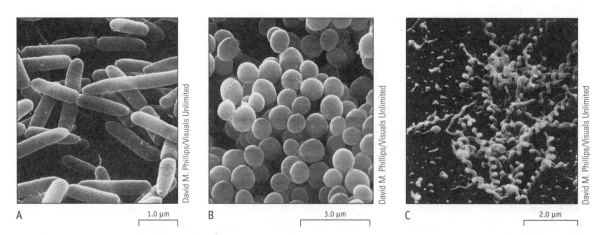

A   1.0 μm

B   3.0 μm

C   2.0 μm

David M. Phillips/Visuals Unlimited

Figure 12-2
Electron Micrographs Showing Typical Shapes of Bacteria. Bacillus: *Salmonella* (A), Coccus: *Micrococcus* (B), and Spirillum: *Spiroplasma* (C).

Look at the prepared slides of each of the three cell types under a compound microscope. If you use oil, clean the slide and the lens with lens paper.

Draw several cells of each type on the data sheet.

Q1. How can you tell the difference between these cells and eukaryotic cells, in terms of size and internal structure?

Two of the most common bacterial **pathogens,** or agents of disease, found in hospitals have the genus names *Streptococcus* (as in *Strep* throat) and *Staphylococcus* (as in *Staph* infections). Examples of these pathogens are shown in Figure 12-3. *Staphylo* means cluster, and *coccus* refers to the way the organism grows. *Staphylococcus* is a coccus-shaped bacterium that grows in clusters. *Strepto* means bent or twisted. *Streptococcus* is a coccus-shaped bacterium that grows in twisted chains.

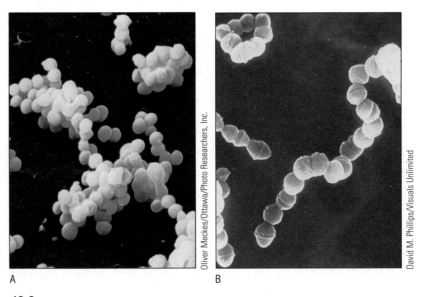

A   B

Figure 12-3
Pathogens. *Staphylococcus,* coccus-shaped bacteria, growing in clusters (A), and *Streptococcus,* coccus-shaped bacteria, growing in twisted chains (B).

*Q2. What would a bacterial genus called **Streptobacillus** look like? Draw a few cells.*

Cyanobacteria

Cyanobacteria are the most ecologically and evolutionarily important photosynthetic bacteria. Some cyanobacteria resemble chloroplasts and may have been predecessors of some types of eukaryotic chloroplasts. Many species fix nitrogen from the atmosphere, making it available to other organisms for protein production. Cyanobacteria may be **unicellular,** or they may exist in **colonies** (groups of cells) or **filaments** (cells arranged in a linear fashion). One type of filamentous cyanobacterium, *Anabaena,* is responsible for the fertility of rice paddies and certain other natural ecosystems. Some cyanobacteria are even motile.

Look at the cyanobacteria specimens available in the lab; draw two or three types. Note whether each one is unicellular, colonial, or filamentous; note whether it is motile. In filamentous types, look for **heterocysts,** thick-walled cells in which nitrogen fixation takes place.

Q3. How can you tell that cyanobacteria are photosynthetic?

Antibiotics/Antibacterials

Most bacteria have a beneficial or neutral role in our lives. Some bacteria in our bodies fight diseases caused by fungi, other bacteria, or other organisms. Some bacteria produce necessary vitamins or help us digest certain foods. Soil bacteria help recycle carbon and other nutrients; some bacteria acquire nitrogen from the air and convert it to a form that can be used by plants and animals to make proteins. Bacteria are used commercially to produce antibiotics, vinegar, and many cheeses and yogurts.

Antibacterials are substances that inhibit or prevent the growth of bacteria. **Antibiotics** are antibacterial substances produced by living organisms. Most commercial antibiotics are forms of the same antibiotics naturally produced by bacteria and fungi to reduce competition.

Humans use antibacterials in attempts to kill pathogens, bacteria that cause disease. Disease is caused by the presence of bacteria or of toxins the bacteria produce. Bacterial diseases include tuberculosis, cholera, anthrax, gonorrhea, and botulism. Antibacterials have helped human populations flourish, but they must be used with caution. Figure 20-5 in the text shows how antibiotics work. Because antibacterials target all prokaryotic cells, both beneficial and pathogenic bacteria are harmed. In addition, killing bacteria that are sensitive to antibacterials encourages the proliferation of antibacterial-resistant bacteria.

Common Household Substances

A variety of products (cleaners, antiseptics, and even mouthwash) claim to have antibacterial properties. In laboratories, bacterial and fungal cultures are grown in Petri dishes that have a layer of a gelatin-like substance called **agar**. Agar is a product of seaweed and is useful for growing cultures because it holds moisture and can be mixed with nutrients that encourage the growth of particular species or groups of species.

For this exercise, Petri dishes with a nutrient medium have been spread with bacterial cultures (*Staphylococcus aureus* or *Escherichia coli*). Small disks soaked in various products (such as bleach, hydrogen peroxide, iodine, mouthwash, alcohol, and antibacterial hand wash) have been placed on the surface of the dishes. If the bacteria cannot grow in the presence of a compound, there will be a clear area (no bacteria) around the disk. This clear area is called a **zone of inhibition** (see Figure 12-4).

Examine the dishes, looking for zones of inhibition.

Q4. Which products were antibacterial?

Q5. Which products seemed to inhibit growth the most?

Q6. Which products were not antibacterial?

Q7. What parts of eukaryotic cells might some antibacterials affect?

Sometimes it is not necessary to use a substance that kills bacteria. For instance, washing your hands with regular soap and water gets your hands clean by the **physical removal** of dirt and bacteria. For many activities, this is sufficient, and it doesn't have the harmful effects that result from overuse of antibacterials.

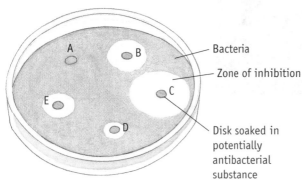

Bacteria

Zone of inhibition

Disk soaked in potentially antibacterial substance

Figure 12-4
Zones of Inhibition. The Petri dish was inoculated with bacteria over its entire surface. Then, disks of paper soaked in potentially antibacterial substances were placed in the dish. After a few days, bacterial growth appeared as shown. Substance C, with the largest zone of inhibition, inhibited bacterial growth the most; Substance A did not inhibit bacterial growth.

Epidemiology

Epidemiology is the study of the incidence and transmission of diseases in populations. When there is an outbreak of an unusual infection, it is the job of an epidemiologist to locate the source of the disease and to learn how to keep that disease from spreading further.

Case Study

During April and May 1998, 11 states reported an increase in cases of *Salmonella* (serotype Agona) infections. *Salmonella* is a bacterium that causes salmonellosis in humans: diarrhea, fever, and abdominal cramps. Symptoms begin 12–72 hours after infection and last between four and seven days. Usually, no antibiotics are required. There are an estimated 2 million to 4 million cases of salmonellosis in the United States each year. However, the Agona strain of *Salmonella* is relatively rare, and 209 people contracted it within two months. It was later found that an additional 39 people in ten states also contracted this type of *Salmonella*.

The state health departments and the Centers for Disease Control and Prevention (CDC) conducted a matched-case study, comparing infected individuals with a control group of healthy household members. In 68 households studied, 56 of the infected people had shopped at a certain supermarket during the three days before illness; 66 percent of the infected people and 36 percent of the controls had consumed a particular brand of toasted oat cereal purchased at the supermarket. Records of average daily consumption of this cereal showed a significant correlation between amount eaten and infection with *Salmonella*.

The CDC cultured an open box of the cereal from the home of a case patient and used gel electrophoresis to positively identify the Agona type *Salmonella*. The same type of bacteria was also found in two samples from unopened boxes of cereal.

All cereal boxes purchased by infected people had a "pull date" of December 1998, implying a production date of April 1998. The cereal company issued a voluntary recall of all cereal from the production line in question, and the outbreak strain of *Salmonella* was found on that line. Because *Salmonella* can stay dormant under dry conditions for a long time, they were not able to positively confirm the original source of infection. A likely suspect is a fortified vitamin spray used on the cereal.

Tracking a Disease in the Classroom

In this exercise, a "disease" will be anonymously spread through the classroom population, and it will be your job to determine where it began.

1. Each student will receive a small dish labeled with an identification number and containing a clear fluid. All but one dish contain water; the remaining dish contains a starch solution. This solution represents the disease you will be tracking. Write your identification number on the data sheet.

2. Put a clean glove on the hand with which you do not write, and liberally smear the liquid in your dish on the palm and fingers of your gloved hand.

3. Now, it is time to come in contact with the population (in this case, your classmates), any of whom may be infected with the disease. Shake hands with three different people. Make sure to smear your gloves together well, and make all your contacts before the liquid on your glove dries out. On the data sheet, write the identification numbers of the people with which you came into contact and the order in which you shook hands.

4. When everyone is done shaking hands, go to the instructor to get tested for the disease. Your instructor will place a drop or two of Lugol's iodine on your glove. If the starch (the disease) is present, it will react with the iodine to form a purplish-black precipitate on the glove; if no disease is present, the iodine will remain yellow.

5. If you are infected, report the identification numbers of the people you contacted, in order, to the class. Fill in the table on the data sheet, and track the infection back to the first infected person by the process of elimination. For instance, if you are infected, but the first person with whom you came in contact is not infected, then you could not have been the origin of the disease.

Q8. Is it possible to determine the first person infected? Why or why not?

Q9. How many people were infected after the first contact?

Q10. How many people could have been infected after the second contact?

Q11. What is the greatest number of people that could have been infected after the third contact?

Q12. How many people were actually infected after three contacts?

Q13. Why might the actual number of infections be smaller than the possible number of infections?

Q14. When an epidemiologist tracks the transmission of a disease in the real world, the job is more difficult than the exercise you just completed. What are three factors that would complicate an epidemiologist's study?

Environmental Plates

Bacteria and other microorganisms (protistans and some fungi) are everywhere. Our immune systems keep us healthy despite constant exposure. To give you an idea of how much is out there, you will inoculate Petri dishes with samples from the classroom. You may take four samples, so develop a hypothesis about the surfaces you will be testing. For instance, maybe one sink in the classroom smells worse than the other. Test both and see if there is a difference in their populations of organisms. Or perhaps the doorknob to the classroom is handled more often than a knob on one of the cupboards. Will their populations of microorganisms differ?

Q15. What is your hypothesis? What is your reasoning? Make sure to take four samples.

Collecting Bacteria

1. Label the bottom of your dish (the side with the agar) with your name and date.

2. Draw two lines across the bottom of the dish, dividing the dish into quadrants (i.e., into 4). Label each quadrant with the surface to be tested.

3. Open a sterile cotton swab. Do not touch the tip. Dip the tip into a tube of sterile saline, and press the tip against the inside of the tube to squeeze out excess saline.

4. Rub the tip on **one** surface: your desktop, a sink faucet, the doorknob, your water bottle, a drinking fountain faucet, a dollar bill, or another surface. Do not touch the tip to anything else.

5. Remove the lid of the Petri dish and gently rub the swab across the surface of the agar in the appropriate quadrant.

6. Repeat Steps 2–5 for three other surfaces, using a new sterile cotton swab for each.

7. Immediately replace the lid and put the dish **upside down** in a designated area; it will be placed in a 37°C incubator (the temperature of the human body).

8. In the next lab period, you can examine your plate for growth.

Q16. Why is it important to replace the lid on your dish immediately?

9. If there is time and a sufficient number of Petri dishes, try a hand washing experiment:

   a. Label the bottom of a plate with your name and date. Draw a line across the bottom center of the plate, labeling one half "unwashed" and the other half "washed."

   b. Remove the lid and touch one finger to the agar on the "unwashed" half of the plate. Immediately replace the lid.

c. Wash your hands vigorously with soap and water.

d. Remove the lid and touch a clean finger to the agar on the "washed" half of the plate. Replace the lid and place the dish upside down in an area; it will be taken to the incubator.

10. **Do not forget to wash your hands after this exercise.**

Examining Your Plates

NOTE: **You will examine your plates during the *next* lab period.**

When you get your plates back, complete this section.

 Do not open your Petri dish. The organisms you put on the dish have multiplied; they may have produced allergens or could be pathogens.

1. Of the different types of microorganisms inoculated onto the dish, only a few have grown: the agar in these Petri dishes contains nutrients that support the growth of certain microorganisms; the temperature of the incubator selects for the microorganisms that can grow at human body temperature.

2. Each bacterial or fungal cell that you applied to the agar has divided to produce many cells that form a **colony**. Use a dissecting microscope to determine how many types of microorganisms are growing in your Petri dish (see Figure 12-5).

    a. **Bacteria** usually form small, round colonies that look somewhat slimy. They are frequently white, off-white, or yellow. Millions of bacterial cells must be present for a colony to be visible to the naked eye.

    b. **Fungi** are discussed in Chapter 21 of the text. To identify fungal colonies, look for spots that appear hairy or fuzzy. They may have produced **spores**, dormant cells that each can form a new colony. Spores will appear as a dark, powdery substance on the colonies or on the surface of the agar.

3. Although you may not be able to identify every colony, you will have an idea of the variety of microorganisms present everywhere. Make sure to look at your classmates' plates, too.

    Q17. Did the results of your test support your hypothesis? If they did not, why do you think they did not?

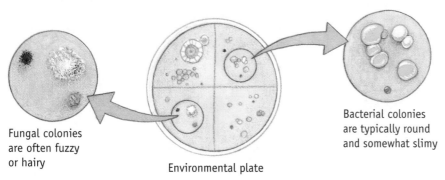

Fungal colonies are often fuzzy or hairy

Environmental plate

Bacterial colonies are typically round and somewhat slimy

Figure 12-5
Identifying Colonies on Your Environmental Plates.

Q18. *Does a type of colony seem to appear on most dishes? Describe the type/appearance of the colony.*

Q19. *Which source showed the greatest diversity of microorganisms? The least diversity?*

When you are done looking at your plates

- Dispose of them in autoclave bags
- Wash your hands thoroughly

Bacteria and Epidemiology
Data Sheet

Bacterial Cell Shape

Draw several cells of each type.

Bacillus	Coccus	Spirillum

Q1. *How can you tell the difference between these cells and eukaryotic cells, in terms of size and internal structure?*

_____

_____

_____

Q2. *What would a bacterial genus called* **Streptobacillus** *look like? Draw a few cells.*

Cyanobacteria

Look at the cyanobacteria specimens available in the lab; draw two or three types. Note whether each is unicellular, colonial, or filamentous and whether it is motile.

Q3. *How can you tell that the cyanobacteria are photosynthetic?* _____

_____

_____

Antibiotics/Antibacterials

Q4. *Which products were antibacterial?* _____

_____

_____

Q5. *Which products seemed to inhibit growth the most?* _____

_____

_____

Q6. *Which products were not antibacterial?* _____

_____

_____

Q7. *What parts of eukaryotic cells might some antibacterials affect?* _____

_____

_____

Epidemiology

Your identification number: _____

Identification numbers of the people with whom you were in contact: _____

    Contact No. 1: _____

    Contact No. 2: _____

    Contact No. 3: _____

IDENTIFICATION NUMBERS OF INFECTED PEOPLE	CONTACT #1	CONTACT #2	CONTACT#3

Q8. *Is it possible to determine the first person infected? Why or why not?* _____

_____

_____

Q9. *How many people were infected after the first contact?* _____

_____

_____

Q10. How many people could have been infected after the second contact? _____

_____

_____

Q11. What is the greatest number of people that could have been infected after the third contact? _____

_____

_____

Q12. How many people were actually infected after three contacts? _____

_____

_____

Q13. Why might the actual number of infections be smaller than the possible number of infections? _____

_____

_____

Q14. When an epidemiologist tracks the transmission of a disease in the real world, the job is more difficult than the exercise you just completed. What are three factors that would complicate an epidemiologist's study? _____

_____

_____

Environmental Plates

Q15. What is your hypothesis about the surfaces you will be testing? What is your reasoning? Make sure to take four samples. _____

_____

_____

Q16. Why is it important to replace the lid on your dish immediately? _____

_____

_____

Q17. *Did the results of your test support your hypothesis? If they did not, why do you think they did not?* _____

_____

_____

Q18. *Does a type of colony seem to appear on most of the dishes? Describe the type/appearance of the colony.* _____

_____

_____

Q19. *Which source showed the greatest diversity of microorganisms? The least diversity?* _____

_____

_____

Questions

In an epidemiological study, the bacterium causing sickness in many people was traced to a deli that purchased a large (40 pound) piece of beef. After thoroughly cooking it, the deli staff placed the whole piece in the refrigerator.

Q20. *Considering the surface area-to-volume ratio of such a large piece of meat, explain why the meat remained warm*

enough to grow bacteria even after it was placed in a refrigerator. How could the deli have avoided this problem?

_____

_____

_____

Q21. *Taking antibiotics makes some people susceptible to fungal infections. Explain why. (You may want to refer to*

"Antibiotics/Antibacterials.") _____

_____

_____

Lab 13   Plant Diversity
This lab accompanies Chapter 22 of *Asking About Life.*

Materials/Equipment

Compound microscopes

Dissecting microscopes

Clean glass slides

Coverslips

Lens paper

Deionized water with pipettes for making wet mounts

Razor blades

Sharps disposal

Live material:

Mosses

Liverworts

Hornworts, if available

One or two stalks of celery

Celery stalk in beaker of water with red food coloring

Ferns

Various fern sori, under dissecting microscopes

Fern gametophytes

Other seedless vascular plants

Pinecones, male and female

Preserved slides of pine pollen (if no fresh male cones)

Pine seeds, with and without seed coats

Flowers to dissect

Various fruits to key

Objectives

1. Become familiar with the major groups of plants

2. Examine some life history stages in the plant groups

3. Learn how plants deal with some challenges of life

4. Understand the importance of the development of vascular tissue and seed production

5. Understand the definition and functions of fruits

Introduction

We are surrounded by plants. Plants allow us to live on the earth; they provide the oxygen we breathe, fibers for clothing, wood for shelter, and they use the energy from sunlight to produce the foods we eat, both plant and animal.

Plants are in the kingdom **Plantae.** All plants are eukaryotic and multicellular. Most are photosynthetic. As living organisms, they face the same challenges of life as all other organisms.

- Plants must have **water.**
- Plants must have an **energy source:** they need access to sunlight. Plants also require carbon dioxide and minerals.
- Plants must **reproduce.** Because most plants are not actively motile, they need a way to **disperse** their young so that parents and progeny are not competing for the same resources.

Figure 13-1
General Plant Life Cycle.
This cycle consists of an alternation of haploid and diploid generations.

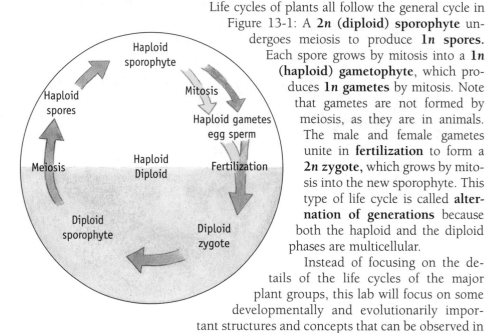

Life cycles of plants all follow the general cycle in Figure 13-1: A **2n (diploid) sporophyte** undergoes meiosis to produce **1n spores.** Each spore grows by mitosis into a **1n (haploid) gametophyte,** which produces **1n gametes** by mitosis. Note that gametes are not formed by meiosis, as they are in animals. The male and female gametes unite in **fertilization** to form a **2n zygote,** which grows by mitosis into the new sporophyte. This type of life cycle is called **alternation of generations** because both the haploid and the diploid phases are multicellular.

Instead of focusing on the details of the life cycles of the major plant groups, this lab will focus on some developmentally and evolutionarily important structures and concepts that can be observed in a lab setting. Life cycle diagrams are provided for reference.

Nonvascular Plants: Bryophytes

Mosses, liverworts, and hornworts are the simplest plants. They are in the division **Bryophyta** (*bryon* = moss, *phyta* = plants) and are referred to as **nonvascular plants** because they do not have special systems for transporting water and nutrients throughout the plant body. Consequently, bryophytes must live in moist environments so that all their body parts are in close contact with water.

Bryophytes produce haploid spores by meiosis. (See Figure 13-2.) Each spore is a single cell surrounded by a protective covering. After the spore is released, the covering breaks open, and the cell inside grows by mitosis into a new plant. The sporophyte, or spore-producing portion of each type of bryophyte, is elevated above the rest of the plant, as shown in Figure 13-3.

Q1. What function does this elevation serve for the plant?

Examine the live specimens on display, noting general characteristics of bryophytes. Distinguish between the gametophyte and/or sporophyte stages of each specimen.

Q2. If bryophytes do not have vascular tissue, what mechanism (that you studied in a prior lab) do they rely upon to transfer materials between cells?

Q3. Notice how small the bryophytes are. What is it about being nonvascular that limits the size of the bryophytes? Why can't they grow large?

Figure 13-2
Life Cycle of a Moss, a Typical Bryophyte.

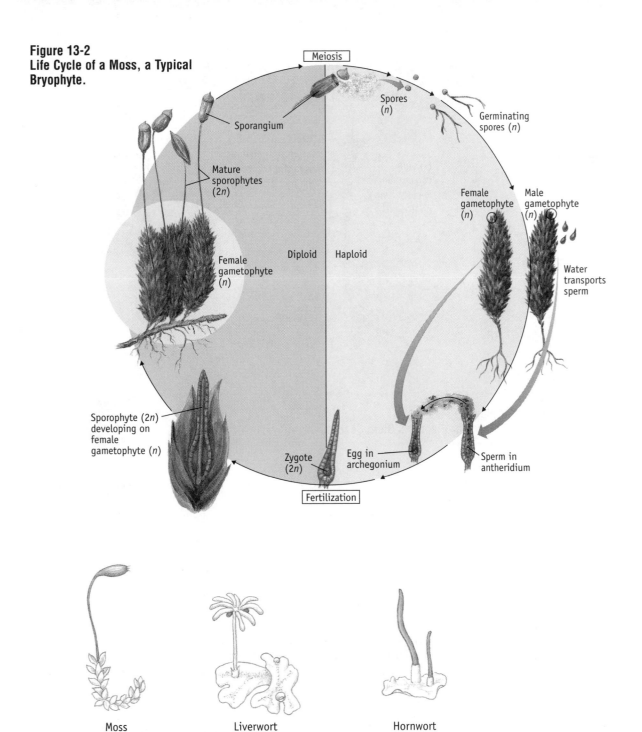

Meiosis

Sporangium

Mature sporophytes (2n)

Spores (n)

Germinating spores (n)

Female gametophyte (n)

Female gametophyte (n)

Male gametophyte (n)

Diploid   Haploid

Water transports sperm

Sporophyte (2n) developing on female gametophyte (n)

Zygote (2n)

Egg in archegonium

Sperm in antheridium

Fertilization

Moss   Liverwort   Hornwort

Figure 13-3
Examples of Bryophytes. Gametophyte and sporophyte stages are shown. Sporophytes are highlighted in green.

Vascular Plants

Grasses, trees, ferns, and shrubs are vascular plants. The leaves, stems, and roots of vascular plants all contain specialized tissues that transport water and nutrients. These tissues are called **xylem** and **phloem** (see Figure 13-4).

Xylem conducts the water and minerals in the soil from the roots up to the leaves. Wood is composed of xylem.

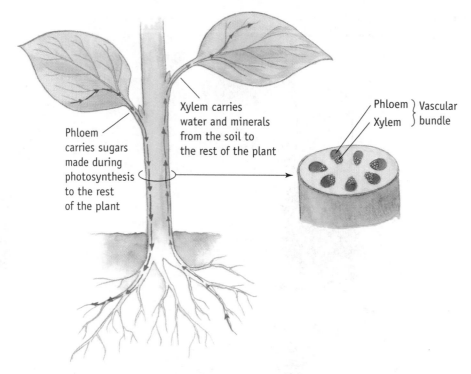

Figure 13-4
Vascular Plant and Stem Section. Xylem and phloem are shown.

Phloem transports sugars and other organic molecules produced in the photosynthetic leaves to the rest of the plant.

Xylem and phloem are typically arranged in groups called **vascular bundles.**

Vascular bundles can be observed in sections of celery.

1. Lay a stalk of celery on a microscope slide and, using a razor blade, make several of the thinnest slices you can. Discard all but two or three of the thinnest slices and add a couple drops of water and a coverslip.

2. Examine your celery under a compound microscope, looking for bundles of vascular tissue. You probably will not be able to distinguish between the xylem and phloem unless you made a very thin section. If you have made a particularly nice section, let your classmates look through your microscope.

3. Draw your celery section, showing the vascular bundles.

   Q4. How does vascular tissue give vascular plants the ability to grow larger than non-vascular plants?

   Q5. If a celery stalk is placed in a container of water with food coloring, the water and food color will visibly move up the stalk toward the leaves. Which type of vascular tissue is responsible for this movement?

Seedless Vascular Plants

Four divisions of vascular plants do not produce seeds: **Pterophyta, Psilophyta, Lycophyta,** and **Equistophyta.** Although the presence of vascular tissue allows these plants to grow much larger than the bryophytes, most seedless vascular plants are limited to fairly wet habitats because all of them produce sperm cells that must swim through water to find and fertilize eggs on another gametophyte.

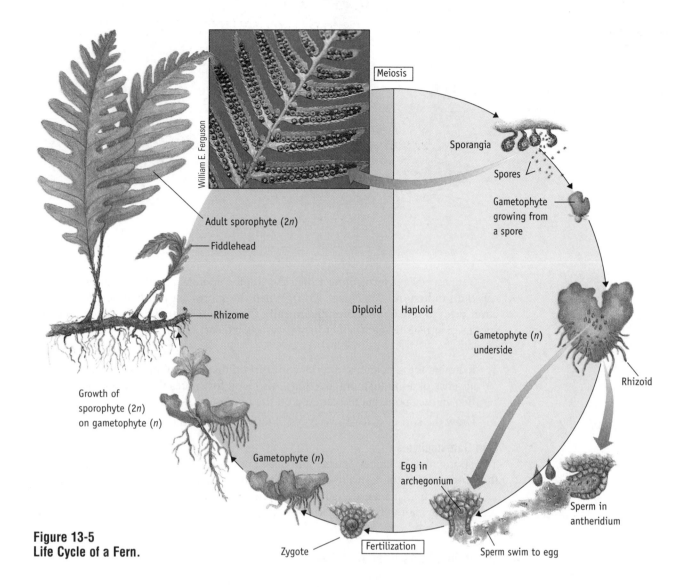

Figure 13-5
Life Cycle of a Fern.

Labels in figure:
Meiosis
Sporangia
Spores
Gametophyte growing from a spore
Adult sporophyte (2n)
Fiddlehead
Rhizome
Diploid
Haploid
Gametophyte (n) underside
Rhizoid
Growth of sporophyte (2n) on gametophyte (n)
Gametophyte (n)
Egg in archegonium
Sperm in antheridium
Zygote
Fertilization
Sperm swim to egg
William E. Ferguson

Ferns

Ferns are in the division **Pterophyta** (*ptero* = feather, *phyta* = plants). The fern life cycle, shown in Figure 13-5, is similar to that of other seedless vascular plants.

1. Examine the fern specimens on display.
2. Look underneath the fern leaves (called **fronds**) and find rusty brown or black regions called **sori** (singular = **sorus**). Sori are the spore-producing regions on ferns, the site of meiosis. The fern fronds are, therefore, part of the fern **sporophyte.** The fronds are upright structures, just as the sporophytes of the nonvascular plants are elevated.

   Q6. What important life function, besides dispersal of spores, do the elevated fronds provide for the plant? Refer to the list in the introduction of this chapter for ideas.

   Q7. In the bryophytes, the gametophyte is the dominant phase of the life cycle. What is the dominant phase of the fern life cycle?

Close Look at Sori

There are several dissecting microscopes set up with sori from various fern species. One sorus is a cluster of spore-producing **sporangia** (singular = **sporangium**). Sporangia are filled with spores. To aid dispersal, some fern species have sporangia on little stalks that catapult the spores out of the sporangia.

Figure 13-6
Sorus. The sorus is the spore-producing region of a fern. It may have an indusium, or covering, over the sporangia, which contain the spores.

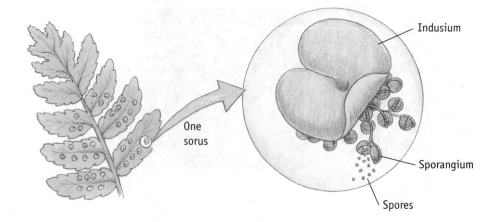

One sorus

Indusium

Sporangium

Spores

Some species of ferns protect their developing sporangia with a covering of tissue called an **indusium** (see Figure 13-6). When the spores are ready to be released, the cover dries to reveal the spores underneath. One way to distinguish between different species of ferns is to see whether they have covered sori, and if so, what shape the covers are.

1. Examine the specimens of ferns with sori on display, looking for the presence or absence of an indusium (covering). Also look for spores that may be present, either in the sporangia or scattered on the microscope stage. Note their small size.

2. Draw the sori of at least two types of ferns, showing the indusium, if present.

Fern Gametophytes

The sporophyte produces spores, and each spore grows into a gametophyte. This is the generation of the fern life cycle that produces gametes (sperm and eggs). In the seedless vascular plants, the gametophyte is small and short-lived. This generation produces the sporophyte. Look at the fern gametophytes: some may have tiny sporophytes growing from them.

Other Seedless Vascular Plants

Examine the other seedless vascular plants on display.

Seed Plants

The evolution of the seed was a big step in the history of plants. A seed consists of a plant **embryo** and a **food source** for that embryo, both surrounded by a protective **seed coat.** The seed coat helps keep the seed from drying out, and the food source gives the young embryo energy to get established before it starts photosynthesizing to produce its own food. If the environment in which a seed lands is inhospitable, the seed can lie dormant—hopefully until the season changes or the weather improves.

The other important development in the evolution of the seed plants was a new method of fertilization. Both bryophytes and seedless vascular plants produce sperm that must swim through water in the environment to reach an egg for fertilization. If the environment is not wet enough at that critical time, the sun shines too hot, or the wind blows, then fertilization does not occur. Consequently, a plant's choice of habitats and windows of opportunity for reproduction are severely limited.

Seed plants do not require free-standing water for sperm to get to eggs; instead, the sperm-producing male gametophyte, in the form of a **pollen grain**, travels to the eggs in the female gametophyte. The pollen grains are carried either by wind or by some kind of pollinator, usually an animal.

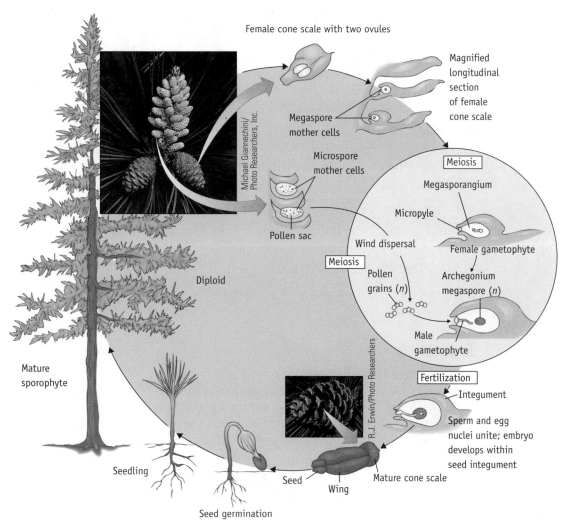

Female cone scale with two ovules

Magnified longitudinal section of female cone scale

Megaspore mother cells

Michael Giannechini/ Photo Researchers, Inc.

Microspore mother cells

Meiosis

Megasporangium

Micropyle

Pollen sac

Wind dispersal

Female gametophyte

Diploid

Meiosis

Pollen grains (n)

Archegonium megaspore (n)

Male gametophyte

Fertilization

R.J. Erwin/Photo Researchers

Integument

Sperm and egg nuclei unite; embryo develops within seed integument

Mature sporophyte

Seedling

Seed

Wing

Mature cone scale

Seed germination

Figure 13-7
Life Cycle of a Pine, a Typical Gymnosperm.

Gymnosperms

Gymnosperms are seed plants without flowers. There are four divisions of gymnosperms: **Coniferophyta, Cycadophyta, Ginkgophyta,** and **Gnetophyta.** This section will focus on the Coniferophyta (which include pines, firs, spruces, cedars, and junipers), although there may be representatives of other divisions in the lab. The life cycle of pines is illustrated in Figure 13-7.

Pinecones

Most conifers produce both male and female cones on the same tree. The male, pollen-producing cones are typically small and produce thousands of pollen grains. Female pinecones are the larger, more familiar pinecones. Each woody scale on a female pinecone bears egg-producing structures.

Look at the pinecones on display.

Pollen

Male cones produce copious amounts of pollen because pine pollen must be carried to the female cones by wind. Heavy pollen production increases the chance that some pollen will reach female cones. To aid dispersal, each pollen grain has small extensions of tissue—like little wings—that keep the pollen grains aloft to be carried farther.

If there are fresh male cones, shake one gently over a drop of water to release some pollen. Add a coverslip, and examine the pollen grains under a compound microscope. If there is no fresh material, examine a prepared slide of pine pollen. Draw a few pollen grains.

> Q8. *Male pinecones are generally located on the tips of the lower branches of the tree, whereas the female cones are located higher in the tree. How might this placement of cones keep the tree from self-fertilization? Why would the tree want to avoid self-fertilization?*

Pine Seeds

Pine seeds are frequently referred to as "pine nuts." They are large enough for you to see all the structures that make a seed a seed.

1. Look at a pine seed still in its "shell," or **seed coat.**

2. Take a pine seed that has been shelled and carefully cut it in half, lengthwise, with a razor blade. In the center is the **embryo.**

> Q9. *What is the tissue surrounding the embryo?*

> Q10. *What will the embryo become?*

3. Draw your pine seed section, labeling the parts.

Angiosperms

Angiosperms Have Flowers

Seed plants that produce flowers are **angiosperms** (see Figure 13-8). Some flowers are wind-pollinated, like the gymnosperms: the male flowers produce pollen, which is carried by the wind to the female flowers, where fertilization takes place. Grasses are a good example of plants that use this type of pollination.

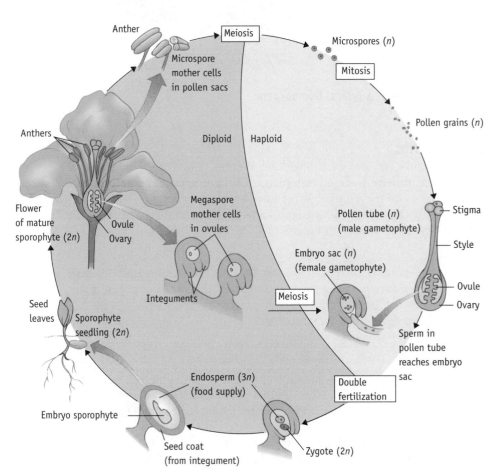

Figure 13-8
Life Cycle of an Angiosperm.

Most angiosperms are pollinated by insects or other animals. The plant can produce much less pollen than wind-pollinated plants because the animal (the **pollinator**) brings the pollen of one flower directly to the **pistil**, or egg-producing structure, of another flower. The flowers on animal-pollinated plants are designed to attract the pollinators with their shapes and colors. Some flowers produce sweet nectar for their pollinators to sip. In many cases, the flowers of a certain plant are designed to attract a specific pollinator.

Q11. Why would it be valuable to a plant to attract a specific pollinator?

Q12. What would happen if every type of flower was attractive to every type of pollinator?

Flower Dissection

Get a flower for observation and dissection. A typical cross section is shown in Figure 13-9.

A flower is made up of **whorls**, or circles, of modified leaves. Usually there are four whorls:

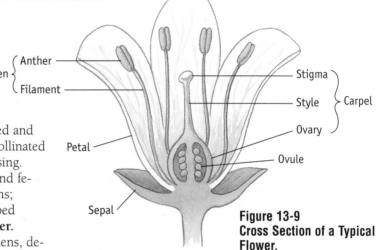

**Figure 13-9
Cross Section of a Typical Flower.**

- **Sepals:** The outermost structures, which are usually green but may be colorful, are the sepals. These enclose and protect the flower when it is developing.
- **Petals:** Petals are typically brightly colored and are used to attract pollinators. In wind-pollinated flowers, the petals are often small or missing.
- **Stamens:** Most flowers have both male and female parts. The male parts are the stamens; they consist of a stalk, the **filament**, topped by the pollen-bearing structure, the **anther**. There may be two stamens or many stamens, depending on the species of flower. If the anthers are mature, you will see the pollen, which is powdery.
- **Carpels (or pistils):** The female parts are called carpels, and there may be one or many of these. Each carpel is made of three parts. The **stigma** is a sticky surface that captures the pollen; the **style** is a tube down which the sperm cell must travel to get to the **ovary**, the site of the egg-containing ovules. After fertilization and development, an ovule becomes a seed and the ovary becomes a fruit.

Draw your flower, labeling all the structures just described. You will need to remove the petals and carefully make a cross section of the ovary to see the ovules inside.

Q13. What is the difference between pollination and fertilization?

Q14. How could it benefit the plant to have both male and female structures in close proximity? How might this situation be a disadvantage?

Angiosperms Produce Fruits

The other difference between gymnosperms and angiosperms is that the angiosperms produce fruits. A fruit encloses and protects the seeds and is the main agent of dispersal of the seeds. Seeds are much larger than spores and cannot be dispersed by the wind like spores can. A fruit may be fleshy and nutritious; in this case, it is eaten by an animal and the seeds pass, unharmed, through the animal's digestive tract to grow. Some fruits are carried by the wind to other locales. Some fruits have projections or bristles that cause them to become caught in the fur (or socks!) of animals.

Q15. What type of dispersal mechanism is used by each of these plants (see Figure 13-10)?

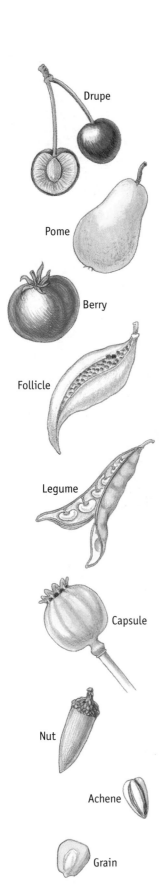

Figure 13-11
Examples of Simple Fruits.

Drupe

Pome

Berry

Follicle

Legume

Capsule

Nut

Achene

Grain

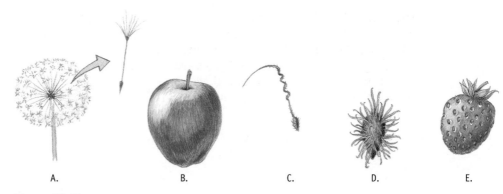

A.   B.   C.   D.   E.

Figure 13-10
Dispersal Mechanism. What dispersal mechanism is used by the plant that produces each of these fruits?

Simple Fruits

Simple fruits are those that develop from a single pistil of one or many fused carpels. If there are several carpels, you will see the separate chambers in the mature fruit—think of the segments of an orange or the chambers inside a tomato when you make a cross section. Each segment or chamber represents one of the fused carpels in the ovary of the flower.

Look at various simple fruits on display and use the key to determine the type of each simple fruit. Make a table showing the common name, fruit type, and dispersal mechanism.

Key to Simple Fruits (see Figure 13-11)

I.   Fruit is fleshy at maturity

    A.   Fruit with one seed that has a stony covering. This covering is part of the old ovary wall . . . . DRUPE

    B.   Fruit with more than one seed

        1.   Fleshy fruit develops from the ovary wall along with other parts of the flower; the ovary wall appears as a "core" around the seeds . . . . . POME

        2.   Fleshy fruit develops from the ovary wall only; the fruit is completely fleshy or pulpy, except for the outer "skin" . . . . BERRY

II. Fruit is dry at maturity

    A.   Fruit splits at maturity

        1.   Fruit in cross section has only one chamber

            a.   Mature fruit splits along one side only . . . . FOLLICLE

            b.   Mature fruit splits along two sides . . . . LEGUME

        2.   Fruit in cross section has more than one cavity . . . . CAPSULE

    B.   Fruit does not open at maturity

        1.   Ovary wall forms a hard, stony covering around the seed; usually one seed . . . . NUT

        2.   Ovary wall does not form a hard, stony covering around the seed

            a.   Seed is attached to ovary wall at one point only; it can be easily separated . . . . ACHENE

            b.   Seed and ovary wall are fused over entire surface; seed is not easily separated . . . . GRAIN

Other Fruit Types

Not all fruits are simple fruits. Fruits that develop from many carpels are either **multiple fruits** or **aggregate fruits.**

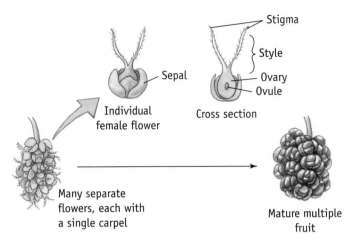

Figure 13-12
Multiple Fruit. A mulberry is formed from many carpels of many flowers.

Multiple fruits are formed from the matured ovaries of many flowers. Examples of multiple fruits are pineapples, mulberries, and figs (see Figure 13-12).

Aggregate fruits are formed from several separate ovaries on one flower. Examples of aggregate fruits are strawberries, blackberries, and raspberries (see Figure 13-13).

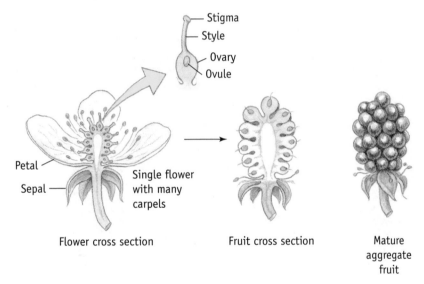

Figure 13-13
Aggregate Fruit. A blackberry is formed from many carpels of a single flower.

Plant Diversity
Data Sheet

Nonvascular Plants: Bryophytes

Q1.   *What function does the elevated sporophyte serve for the plant?* _____

_____

_____

Q2.   *If bryophytes do not have vascular tissue, what mechanism (that you studied in a prior lab) do they rely upon to transfer materials between cells?* _____

_____

_____

Q3.   *Notice how small the bryophytes are. What is it about being nonvascular that limits the size of the bryophytes? Why can't they grow large?* _____

_____

_____

Vascular Plants
Draw your celery section, showing vascular bundles.

Q4. *How does vascular tissue give vascular plants the ability to grow larger than nonvascular plants?* _____

_____

_____

Q5. *If a celery stalk is placed in a container of water with food coloring, the water and food color will visibly move up the*

stalk toward the leaves. Which type of vascular tissue is responsible for this movement? _____

_____

_____

Ferns

Q6. *What important life function, besides the dispersal of spores, do the elevated fronds provide for the plant? Refer to the*

list in the introduction of this chapter for ideas. _____

_____

_____

Q7. *In bryophytes, the gametophyte is the dominant phase of the life cycle. What is the dominant phase of the fern life*

cycle? _____

Closer Look at Sori
Draw the sori of at least two types of ferns, showing the covering, if present.

Pinecones

Draw a few pollen grains.

Q8. *Male pinecones are generally located on the tips of the lower branches of the tree, whereas the female cones are located higher in the tree. How might this placement of cones keep the tree from self-fertilization? Why would the tree want to avoid self-fertilization?* _____

_____

_____

Pine Seeds

Q9. *What is the tissue surrounding the embryo of the pine seed?* _____

_____

_____

Q10. *What will the embryo become?* _____

_____

_____

Draw your pine seed section, labeling the parts.

Angiosperms Have Flowers

Q11. Why would it be valuable to a plant to attract a specific pollinator? _____

_____

_____

Q12. What would happen if every type of flower was attractive to every type of pollinator? _____

_____

_____

Draw your flower, labeling the whorls (sepal, petal, stamen, and carpel), filament, anther, stigma, style, ovary, and ovule. You will need to remove the petals and carefully make a cross section of the ovary to see the ovules inside.

Q13. What is the difference between pollination and fertilization? _____

_____

_____

Q14. How could it benefit a plant to have both male and female structures in close proximity? How might this situation be

a disadvantage? _____

_____

_____

Angiosperms Produce Fruits

Q15. *What type of dispersal mechanism is used by the plant that produces each fruit shown in Figure 13-10?*

A. _____

B. _____

C. _____

D. _____

E. _____

Simple Fruits

Make a table showing the common name, fruit type, and dispersal mechanism used by each fruit you can identify using the key.

Questions

Q16. If a fruit, by definition, develops from a flower and encloses seeds, then many common produce items that we call vegetables are really fruits. List four. _____

_____

_____

Q17. What, then, are true vegetables? List four. _____

_____

_____

Q18. Many people have allergies to pollen. Pollen from two groups of plants discussed in this lab is likely to cause such allergies. Which groups of plants? Why? _____

_____

_____

Q19. What is the advantage of having pollen instead of free-swimming sperm? _____

_____

_____

Q20. Why is it important to have a mechanism for dispersing young? What would happen if a plant dropped all of its spores or seeds at its own base? _____

_____

_____

Q21. List at least three ways in which the development of vascular tissue and seeds has allowed the gymnosperms and angiosperms to become successful in many types of environments. _____

_____

_____

Lab 14   Animal Diversity
This lab accompanies Chapters 23 and 24 of *Asking About Life.*

Objectives

1. Realize the diversity of animals

2. Understand adaptations to environments by different animal groups, especially with regard to feeding and locomotion

3. Understand how different types of animals fulfill their requirements for life

4. Compare the lifestyle of a free-living organism with that of a parasite

Introduction

When you go to the zoo, you see an array of animals: elephants, bears, giraffes, ostriches, hippos, lions, monkeys, and perhaps even snakes and lizards. However, as far as animal diversity goes, this collection represents a miniscule sampling. All common zoo animals are **vertebrates** (animals with backbones), which comprise only 1% of all animal species. The other 99% of animal species are **invertebrates,** animals without backbones. Invertebrates include insects, snails, squid, and worms.

An **animal** is a eukaryotic, multicellular, heterotrophic organism that develops from an embryo. Most animals reproduce sexually and have nervous tissue and muscle tissue. Like all other living organisms, animals need a source of **water** and **energy,** they need to **reproduce,** and they need to **disperse.** Animals employ a variety of methods to accomplish these tasks. You will examine some of these strategies after some notes on classification.

Classification

Between 1 million and 2 million animal species have been classified; another 2 million to 28 million species have not yet been discovered or described. Animals in the subkingdom **Eumetazoa** are classified according to their mode of development and by their symmetry. Sponges, in the subkingdom **Parazoa,** have neither organs nor symmetry.

Development

Development will be discussed thoroughly in Lab 15. After fertilization, an animal zygote divides by mitosis to form a ball of cells, as shown in Figure 14-1. This ball folds in on itself to form either two or three cell layers, depending on the species. If the location of the infolding becomes the mouth of the animal, that animal is called a **protostome** (*proto* = first, *stoma* = mouth); if the mouth is formed secondarily, the animal is called a **deuterostome** (*deuteros* = second).

If the embryo has two cell layers (endoderm and ectoderm), the animal is an **acoelomate** (*a* = without, *koilos* = hollow). An embryo with three cell layers (endoderm, mesoderm, and ectoderm) is a **coelomate** if it has an internal body cavity lined with mesoderm; it is a **pseudocoelomate** (*pseudo* = false) if it has an internal body cavity not lined with mesoderm. See Figure 23-7 in the text.

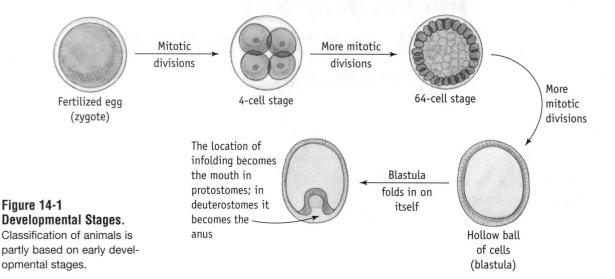

**Figure 14-1
Developmental Stages.**
Classification of animals is partly based on early developmental stages.

Fertilized egg (zygote) → Mitotic divisions → 4-cell stage → More mitotic divisions → 64-cell stage → More mitotic divisions → Hollow ball of cells (blastula) → Blastula folds in on itself → The location of infolding becomes the mouth in protostomes; in deuterostomes it becomes the anus

Symmetry

Another characteristic used to classify the Eumetazoa is symmetry. Some animals have **radial symmetry**: their appearance does not change as they are rotated along a central axis. See Figure 14-2. Jellyfish have radial symmetry, as shown in Figure 14-2. Radially symmetric animals have a top surface and bottom surface but no front, back, left, or right.

**Figure 14-2
Types of Symmetry.** A jellyfish (A) exhibits radial symmetry: its appearance does not change as it is rotated along a central axis. A scorpion (B) exhibits bilateral symmetry: it can be divided into right and left sides that are mirror images of each other. A bilaterally symmetric animal, such as a poison arrow frog (C), has a defined anterior, posterior, dorsal side, and ventral side.

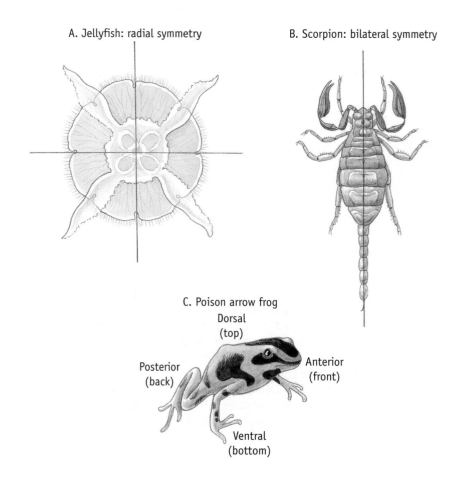

A. Jellyfish: radial symmetry

B. Scorpion: bilateral symmetry

C. Poison arrow frog
Dorsal (top)
Posterior (back)
Anterior (front)
Ventral (bottom)

Other animals have **bilateral symmetry:** they may be divided into right and left sides that are mirror images of each other. Lizards and humans exhibit bilateral symmetry. These animals have a defined **anterior** (front) and **posterior** (back) as well as a **dorsal** side (top) and **ventral** side (bottom). Having a "front" usually means there is a preferred direction of movement, "head first." Because the head encounters new stimuli first, the head includes sense organs.

Q1. What three types of stimuli should the head be able to detect?

Animals in Their Environments

In this lab, you will look at animals in two different habitats: aquatic and terrestrial. Animals from several major phyla will be described with respect to their adaptations to their environments, focusing on feeding and locomotion. Read the descriptions and answer the questions. In addition, examine any live or preserved specimens provided.

Note that although many of these animals appear to have limitations in size, body form, sensory adaptation, motility, or a combination of these, each animal is successful in its environment.

Habitat 1: Rocky Intertidal

One environment with the richest species diversity is the rocky intertidal zone (see Figure 14-3). The autotrophs in this habitat are seaweeds (large marine algae), microscopic marine algae, and other protists. Besides containing microscopic protists, seawater is filled with minerals, organic debris, and tiny larval forms of many different marine organisms. Many animals filter these nutrients out of water.

The water level in the intertidal zone is constantly changing with the tides: the middle range of this zone is submerged twice a day and exposed twice a day.

Besides experiencing wet and dry conditions, organisms that live in the intertidal zone are submitted to pounding waves. Many animals resist being washed away by attaching themselves firmly to rocks.

Q2. How would being attached to one spot make it difficult to obtain food and water or to reproduce and disperse?

A. This is a **sponge,** in the phylum Porifera (see Figure 14-4). It is **sessile,** permanently attached to a rock. As a member of the subkingdom Parazoa, sponges have no well-defined symmetry and no organs. Because it cannot move toward its food, it brings in food by creating a flow of water through its body. Special cells filter food particles out of the water. See Figure 23-8 in the text.

B. These are **sea anemones,** in the phylum Cnidaria. See Figure 14-5 (A).

Q3. What type of symmetry do these animals have?

Like sponges, these anemones are sessile. They are green because of the symbiotic algae that live within their tissues. The algae provide an anemone with some nutrients. For the rest of its food, the anemone catches prey with its tentacles. It is armed with stinging cells called **cnidocytes** that can immobilize a crab, shellfish, or fish. See Figure 14-5 (B). The tentacles then transfer the prey to the anemone's mouth and into its "stomach," the **gastrovascular cavity** in which the prey is bathed with digestive juices. Any hard or indigestible parts are expelled through the mouth. Cnidarians do not have an anus, so the mouth functions as a mouth and an anus.

Because an anemone is not actively motile, it cannot escape predators or the drying conditions of low tide. When the water level lowers, or if the anemone is threatened by a predator, the anemone will close up, pulling in its tentacles.

Figure 14-3
Rocky Intertidal Zone.
Some of the common organisms found there are shown. Different organisms live at different vertical heights within the habitat. The high- and low-tide lines mark the highest and lowest levels the water reaches.

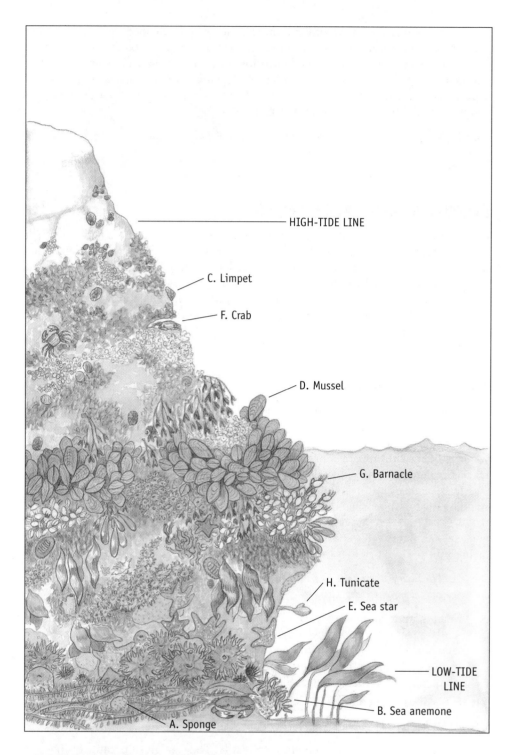

HIGH-TIDE LINE

C. Limpet

F. Crab

D. Mussel

G. Barnacle

H. Tunicate

E. Sea star

LOW-TIDE LINE

B. Sea anemone

A. Sponge

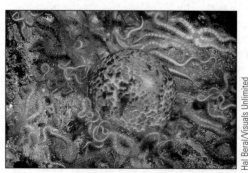

Hal Beral/Visuals Unlimited

Figure 14-4
Tethya aurantia. This orange puffball sponge is surrounded by brittle stars from the phylum Echinodermata. The sponge filters sea water for food particles by drawing water in and then forcing it out through the many pores on the sponge's surface.

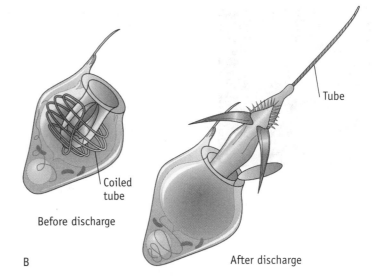

Tube

Coiled tube

Before discharge

After discharge

B

Figure 14-5

A. Giant Green Anemones *(Anthopleura xanthogrammica).* A ring of green tentacles surrounds the mouth.

B. Cnidocyte Before and After Discharge. When triggered, a tiny barbed harpoon shoots out of the cell. If it enters the flesh of another animal, it injects a toxin that paralyzes the victim so that the anemone can pull it into its mouth.

Q4. How does this behavior protect the anemone from water loss?

Q5. What functions might the rocks attached to the outside of the anemone serve?

C. This **limpet** is in the phylum Mollusca. It moves around on an adhesive **foot** in search of algae growing on rocks. Limpets use a **radula,** a ridged structure, to scrape algae off the rocks (see Figure 14-6). When the tide lowers, many limpets stay above the water line, sealing the edges of their shells against the rocks to keep in moisture. When the water level rises, the limpets resume feeding.

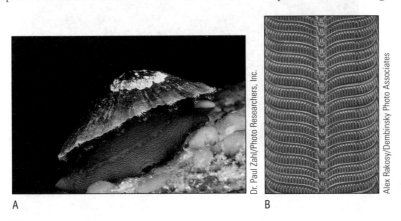

A

B

Figure 14-6

Limpet. Side view (A) of a limpet. Its head bears two sensory tentacles. A typical radula (B) is used to scrape algae from rock surfaces. In some mollusks, it may be adapted for catching prey, consuming meat, or both.

Q6. From what might the limpets be protecting themselves when they stay above the water level?

D. This **mussel** is also in the phylum Mollusca (see Figure 14-7). It has two shells that can be closed tightly to protect the mussel from water loss and predation. Instead of moving around on its foot, a mussel uses its foot to form **byssal threads,** which tether the mussel to the rocks. Like sponges, mussels create a flow of seawater through their bodies, and they filter small particles of food from the water.

Figure 14-7
Mussels (Mytilus californianus) Anchored to a Rock.
The mussels are surrounded by two species of barnacles. Notice that some barnacles are attached to mussels. Barnacle larvae settled on the mussels and grew to adulthood there, a testament to the stability of the mussels' attachment to the rock.

Gerald and Buff Corsi/Visuals Unlimited

Q7. A mussel exhibits bilateral symmetry but has no defined head. Does it need one? Explain.

E. **Sea stars** are in the phylum Echinodermata (see Figure 14-8). The undersurfaces of a sea star's five flexible arms are covered with **tube feet:** tiny extensible structures with tips like suction cups. The sea star can move along surfaces and still be firmly attached. The tube feet are also used to pull open shellfish such as mussels (see Figure 14-9). When the mussel shell has been opened a bit, the sea star pushes its stomach through its mouth and into the opening in the mussel shell. The stomach secretes digestive enzymes and digests the mussel's soft body. Sea stars eat members of almost all animal phyla.

Q8. Why do many marine biologists prefer the term "sea star" to the common term "starfish"?

Sea stars prey heavily on mussels but cannot tolerate being out of the water as long as mussels can. Mussels live as high in the intertidal zone as they can to avoid predation by sea stars.

Q9. What are two life requirements that mussels compromise to avoid predation?

Q10. Sea stars do not have eyes to locate their prey. What sense must they use to perform this function?

Terry Donnelly/Dembinsky Photo Associates

Figure 14-8
Sea Stars (Pisaster ochraceous). Near a typical food source: mussels.

Figure 14-9
Sea Star. A sea star uses its tube feet to hold onto and pull open a mussel (A). The shell needs to open only slightly to allow the sea star to push its stomach out its mouth and into the mussel's shell. If you could see through the sea star (B), you would see how the stomach comes out the sea star's mouth and enters the mussel's shell to digest the mussel.

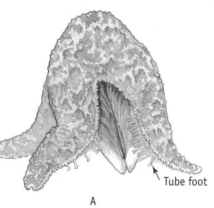

Tube foot

A

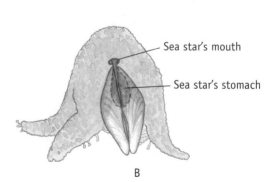

Sea star's mouth

Sea star's stomach

B

F. This shore **crab** is in the phylum Arthropoda (see Figure 14-10). It moves actively on its five pairs of legs. The first pair of legs is modified into claws that the crab uses to pick up small pieces of algae or bits of organic debris. The claws are also used to pick apart any dead animals the crab may encounter. Larger species of crabs can use their claws to capture prey.

Q11. *What kind of symmetry does the crab exhibit?*

Q12. *How could a shore crab resist being washed to sea by waves?*

G. **Barnacles** are also arthropods. Unlike crabs, shrimp, insects, and most other arthropods, barnacles are sessile. They secrete calcium-containing plates that cement them to rocks, boats, pilings, or even whales. Only the legs of a barnacle ever emerge from the calcium plates. A barnacle uses its legs to filter food particles out of the seawater and to pass those particles into its mouth (see Figure 14-11).

Because they cannot move around to find mates, many sessile marine organisms reproduce sexually by broadcasting sperm and eggs into the seawater. The downside to this method of reproduction is that a great number of gametes must be produced to ensure they will encounter each other (similar to the wind-borne pollen of pine trees).

Barnacles avoid broadcasting gametes by having extraordinarily long penises. A barnacle can extend its penis several centimeters and insert it into the chamber formed by the calcium plates of another barnacle. The second barnacle protects the fertilized eggs until the swimming larvae hatch.

Q13. *What is a disadvantage of barnacles' reproduction method, in terms of maintaining genetic diversity?*

Q14. *What life cycle stage disperses in barnacles? How is genetic diversity affected?*

Q15. *What animals in this section probably reproduce by broadcasting their gametes into the sea? Explain your choices.*

H. This **tunicate** (see Figure 14-12) is in the phylum Chordata—the same phylum to which humans belong. In its larval stage, a tunicate shares characteristics with other chordates, but these features are lost upon growth into an adult. These features are discussed in Lab 15.

Tunicates are sessile and, like sponges and mussels, they filter feed by drawing currents of seawater through their bodies.

Fertilization, for all animals listed in this section, results in the formation of a swimming larva, which feeds and is carried with the ocean currents (see Figure 14-13). When the time is right and the larva finds the appropriate location (often near others of its species), it settles on a substrate and grows into an adult.

Q16. *Could a marine animal survive in fresh water? Explain your answer.*

Figure 14-10
Lined Shore Crab *(Pachygrapsus crassipes).* Notice the crab in the background, hiding in a crack in the rock.

Figure 14-11
Gooseneck Barnacles *(Lepas anserifera).* Their feathery legs are extended for feeding. Compare these legs with those of another arthropod, the crab.

Figure 14-12
Tunicate. This tunicate takes in water through the opening on its top, filters food from the water by trapping it in sticky mucus, then expels the water and its waste products through the opening on its side.

A. Crab larvae.

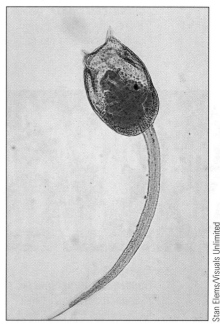

D. Tunicate larva.

(Barnacle larva image — William C. Jorgenson/Visuals Unlimited)

B. Barnacle larva.

(Sea star larva image — John D. Cunningham/Visuals Unlimited)

C. Sea star larva.

(Mussel larva image — Wim Van Egmond)

E. Mussel larva.

Figure 14-13 Larval Stages of a Variety of Marine Organisms. All swim and feed in the open ocean and serve a role in dispersal.

For humans to extract energy from food by cellular respiration, large pieces of food must first be broken into smaller pieces that can fit in the mouth. These pieces are then chewed into smaller pieces that make their way into the stomach for further digestion.

> *Q17. Of the animals examined so far, only anemones, sea stars, and crabs can eat large food. None of these animals have jaws with which to tear apart large food, so how does each of these three animals handle large food to make it suitable for digestion?*

Habitat 2: Urban Park

An urban park presents challenges for an animal different from those of the rocky intertidal zone, but the requirements for food, water, reproduction, and dispersal are the same (see Figure 14-14). A terrestrial environment does not have the ocean dangers of being battered or ripped away, but food is not as plentiful as it is in the intertidal zone. Even a freshwater stream or lake has only a small amount of organic matter suspended in the water. Animals in a terrestrial or freshwater environment generally must move around to find food.

Motile animals do not have the same obstructions to dispersal and reproduction that sessile marine animals have. However, living in a terrestrial habitat also means

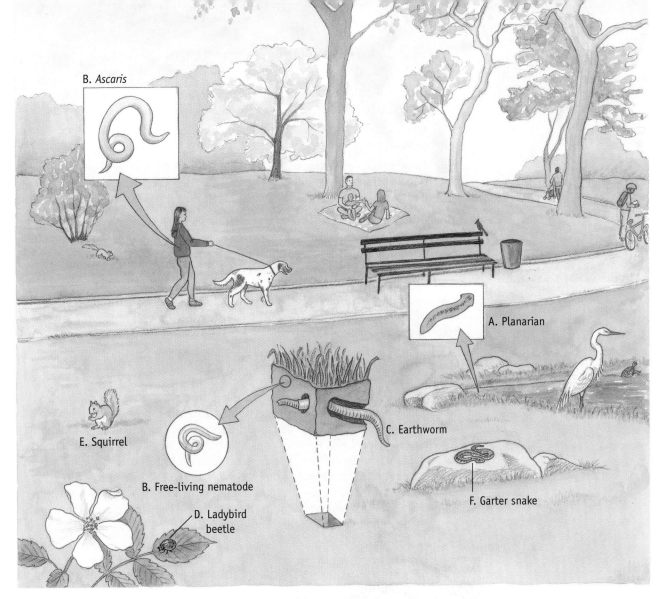

B. *Ascaris*

A. Planarian

E. Squirrel

B. Free-living nematode

D. Ladybird beetle

C. Earthworm

F. Garter snake

Figure 14-14
Urban Park. Some common organisms found there: The environmental conditions are different from those in the rocky intertidal zone, and the animals have adaptations to meet those conditions.

that gametes cannot be merely thrown into the environment. All completely terrestrial animals (and some aquatic animals) have some mode of **internal fertilization**: the male inserts sperm into the female's body. This method greatly increases the chances of fertilization and requires the production of significantly fewer gametes.

Another major difference between completely terrestrial animals and aquatic animals is that the terrestrial animals must have adaptations to conserve water.

Q18. What would happen to the water in terrestrial animals' bodies if they had no mechanisms for conservation?

A. **Planarians** are in the phylum Platyhelminthes (flatworms). See Figure 14-15. As their common name implies, planarians are extremely flat—a fraction of a millimeter thick. They live in freshwater ponds and streams and use cilia to glide across a film of mucus they secrete.

The mouth of a planarian is at the end of a muscular pharynx that extends from the ventral surface. Digestive enzymes are spilled onto the prey, either smaller animals or dead organisms, and the pharynx sucks small pieces of food into the body, where digestion continues.

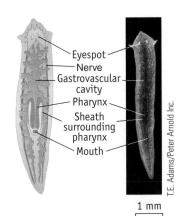

Eyespot
Nerve
Gastrovascular cavity
Pharynx
Sheath surrounding pharynx
Mouth

1 mm

T.E. Adams/Peter Arnold Inc.

Figure 14-15
Planarian (*Dugesia dorotocephala*). Eyespots and major internal structures are shown.

Planarians have **eyespots** that can detect light but do not form images. A planarian will actively move away from light.

Q19. *Why do you think it is important for a planarian to be able to detect and move away from a light source? Could light be dangerous to a flatworm? Explain.*

Q20. *Because their eyesight is not well developed, what sense must planarians possess to locate prey?*

Q21. *Planarians have no organs specialized for respiration or circulation. How do they distribute oxygen and nutrients to their cells? How does this relate to their flattened appearance?*

B. **Nematodes** (roundworms) are in the phylum Nematoda. A nematode has only longitudinal muscles—that is, muscles that run the length of its body. It can only make thrashing, back-and-forth movements. A nematode's body is covered with a tough protective layer called a **cuticle.**

The small nematode is a free-living species in the soil. A fistful of soil contains millions of microscopic nematodes. This nematode ingests organic material in the soil—it is important in decomposition and nutrient cycling (see Figure 14-16).

Q22. *Would you expect these organisms to have specialized organs for respiration and circulation? Why or why not?*

Although most nematode species are free-living, there are parasitic nematodes that live in or on nearly all vertebrates, many invertebrates, and almost every type of plant. The large nematode, *Ascaris,* is a parasite in the human intestine (see Figure 14-17). Every day, 200,000 eggs are laid by a female and leave the host's body in feces. The eggs lie dormant in soil or water until a new host becomes infected, usually from hand to mouth. The immature worms migrate through the body until they arrive in the intestine, where they grow into adults. The adults usually remain in the same host. They feed by sucking up the partially digested contents of the host's intestines.

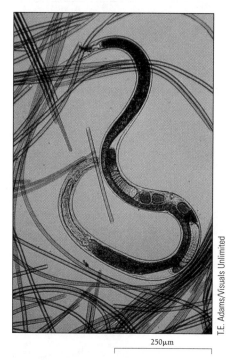

250μm

T.E. Adams/Visuals Unlimited

Figure 14-16
Free-Living Nematode. The nematode eats the surrounding cyanobacteria.

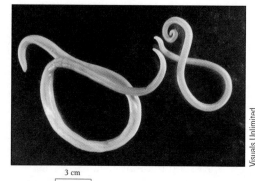

3 cm

Visuals Unlimited

Figure 14-17
Ascaris lumbricoides. The female is the larger individual; the male is smaller.

Q23. *A parasitic intestinal nematode is bathed in the digestive enzymes of its host. How would you expect the cuticle of this worm to compare with the cuticle of a free-living nematode?*

Q24. *All nematodes have sense organs in their lips, but chemical receptors are found only in free-living nematodes, not in parasitic nematodes. Explain why this difference exists, based on how each type of worm lives.*

Q25. **Ascaris** *infection is typically found in the tropics, but some communities in North America are infected. Based on the method of transmission, what kind of living conditions do infected populations have? How could you avoid infection in such circumstances?*

C. This **earthworm** is in the phylum Annelida. *Annelida* means little rings and refers to the segmentation of the worm's body. Each segment has four pairs of **setae,** small bristles that provide traction as the worm burrows through soil (see Figure 14-18). Each segment can elongate and contract, giving the worm a great range of movement.

The mouth of an earthworm functions to swallow particles of soil, which are then pumped through the worm's body. Organic material is digested; indigestible materials are passed through the worm's anus as castings. The combination of their burrowing activity and their castings keeps soil mixed and aerated.

**Figure 14-18
Electron Micrograph of the Common Earthworm (Lumbricus terrestris).** The setae can be seen on each segment, and the mouth is at the tip of the head (center).

Q26. *One of the few times earthworms come above ground is when the soil is soaked with water. What do they need above the ground? For what important process is this required?*

Q27. *How are earthworms ill suited for being above ground? Devise two problems earthworms would encounter on the soil's surface.*

Q28. *There are many types of marine annelids. Some of these are filter feeders; some are active predators. How would each of these types of worms differ from earthworms? Consider mobility, mouthparts, and sensory structures.*

D. **Beetles** are in the phylum Arthropoda. This ladybird beetle can move rapidly on its three pairs of legs. It also has two pairs of wings underneath protective wing covers: beetles can fly.

The hard **exoskeleton,** which covers all exposed body parts, supports the beetle's body, giving it structure and a place for muscles to attach. The exoskeleton also protects the body from injury and water loss in a terrestrial environment.

Ladybird beetles are predators of aphids, another type of arthropod shown in Figure 14-19. Aphids have piercing, sucking mouthparts that they use to puncture plant stems or leaves and suck out sugars. Ladybird beetles have mouthparts suitable for biting and chewing.

Ladybird beetles and other insects have small openings in their exoskeleton called **spiracles.** Inside the beetle, the spiracles branch into small tubes that reach all cells of the body.

Q29. *What do you suppose these spiracles and branching tubes are for? Why do insects need them?*

**Figure 14-19
Ladybird Beetle (Coccinella septumpunctata) Feeding on Aphids.**

Figure 14-20
Grey Squirrel (Sciurus carolinensis). This animal can use its forelimbs for manipulating food, grooming, and climbing.

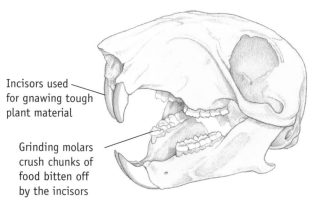

Incisors used for gnawing tough plant material

Grinding molars crush chunks of food bitten off by the incisors

Figure 14-21
Skull of Eastern Grey Squirrel. Herbivores must chew their food into small pieces because plant material is difficult to digest.

E. A **squirrel** is a vertebrate, in the phylum Chordata. Its skin protects against water loss and is covered with an insulating layer of fur. As an **endotherm**, a squirrel captures the heat produced by cellular respiration and uses it to help maintain a constant body temperature.

The squirrel's four legs are used for running and climbing. The front legs can grasp nuts, fruits, and seeds (see Figure 14-20). Some of these items are tough, so the squirrel's jaws and teeth must be very strong. The front teeth grow continuously throughout the squirrel's life, so the squirrel must keep gnawing on food or tree branches (and sometimes on the wood of houses!) to limit the length of their teeth and to keep their teeth sharp and clean. The front layers of a squirrel's front teeth are harder than the back part, so the teeth are self-sharpening. The squirrel's flattish molars crush pieces of food before they are swallowed (see Figure 14-21).

F. A **garter snake** is another vertebrate (see Figure 14-22). Its skin is covered with a layer of scales to help conserve water. As an **ectotherm**, the snake receives body heat from the sun. If an ectotherm becomes too cold, its enzymes cannot function at a practical rate: muscles cannot respond quickly, senses are dulled, digestion barely takes place, and so on.

Q30. What could happen to an ectotherm that gets too hot?

Snakes have no legs, but they are predators. Garter snakes eat worms, insects, fishes, frogs, tadpoles, small mammals, and sometimes even small birds. How could a legless creature catch prey? A snake usually moves by undulating, pushing curves of its body against objects and rough surfaces on the ground. Snakes are sensitive to vibrations in the ground and can see movement with their eyes. The sensitive chemical sensors of a snake are located on the roof of its mouth: a snake flicks its tongue in and out to fan odors over these sensors. Once the garter snake locates prey, it can lunge its head, mouth open, to seize the prey. Unlike some snakes that have fangs to deliver a toxin, a garter snake has many small, sharp teeth. Because a snake cannot chew or tear the prey apart, it holds it firmly in its mouth and swallows the prey whole.

To eat prey larger than the snake's head, the snake's lower jaw is hinged to the skull by an elastic ligament that permits the mouth to open wide (see Figure 14-23). In turn, each side of the jaw moves slightly forward and back until the prey is swallowed.

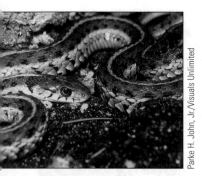

Figure 14-22
Eastern Garter Snake (Thamnophis sirtalis). Notice the snake's protective coloration.

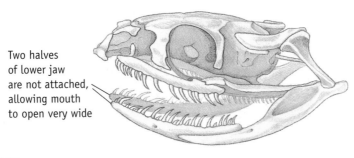

Two halves of lower jaw are not attached, allowing mouth to open very wide

Figure 14-23
Garter Snake Skull. Because meat is easy to digest, a carnivore can ingest any piece that will fit into its mouth. A garter snake swallows its prey whole: the backward-pointing teeth are adapted for catching, gripping, and directing prey toward the stomach.

Q31. *How can you tell, just by looking at the jaws, that the squirrel is an herbivore and the snake is a carnivore?*

Q32. *Why is it impractical for snakes to have fur? Think about how they move and whether they need insulation.*

Q33. **Therm** = *heat,* **endo** = *within,* **ecto** = *outside: explain why* **endotherms** *and* **ectotherms** *are given these names.*

Examine any other live or preserved animal specimens available for you in the lab. By observation and the information you learned in the last exercise, consider how each specimen is adapted for its environment, what kind of locomotion it employs, and how it feeds. Record your observations on the data sheet.

Phyla Represented in This Lab

The names of phyla are in bold print.

I. Protostomes

   A. Parazoa

      1. **Porifera** (sponges)

   B. Eumetazoa

      1. Acoelomates

         a. **Cnidaria** (jellyfish, sea anemones, corals, hydroids, and cubozoans)

         b. **Platyhelminthes** (flatworms)

      2. Pseudocoelomates

         a. **Nematoda** (roundworms)

      3. Coelomates

         a. **Arthropoda** (insects, spiders, crabs, shrimp, barnacles, and so on)

         b. **Annelida** (segmented worms)

         c. **Mollusca** (snails, squid, octopuses, clams, nautiloids, and so on)

II. Deuterostomes

   A. Eumetazoa

      1. Coelomates

         a. **Echinodermata** (sea stars, sea urchins, brittle stars, sea cucumbers, and so on)

         b. **Chordata** (tunicates, *Amphioxus,* and vertebrates)

Animal Diversity
Data Sheet

Classification
Symmetry

Q1. What three types of stimuli should the head of an animal with bilateral symmetry be able to detect? _____

_____

_____

Animals in Their Environments
Habitat 1: Rocky Intertidal

Q2. How would being attached to one spot make it difficult to obtain food or water or to reproduce and disperse?

_____

_____

_____

Q3. What type of symmetry do sea anemones have? _____

Q4. How does an anemone's ability to close up, pulling in its tentacles, protect it from water loss? _____

_____

_____

Q5. What functions might the rocks attached to the outside of an anemone serve? _____

_____

_____

Q6. From what might limpets be protecting themselves when they stay above the water level? _____

_____

_____

Q7. A mussel exhibits bilateral symmetry but has no defined head. Does it need one? Explain. _____

_____

_____

Q8. Why do many marine biologists prefer the term "sea star" to the common term "starfish"? _____

_____

_____

Sea stars prey heavily on mussels but cannot tolerate being out of the water as long as mussels can. Mussels live as high in the intertidal zone as they can to avoid predation by sea stars.

Q9. What are two life requirements that mussels compromise to avoid predation? _____

_____

_____

Q10. Sea stars do not have eyes to locate their prey. What sense must they use to perform this function? _____

Q11. What kind of symmetry does a crab exhibit? _____

Q12. How could a shore crab resist being washed out to sea by waves? _____

_____

_____

Q13. What is a disadvantage of barnacles' reproduction method, in terms of maintaining genetic diversity? _____

_____

_____

Q14. What life cycle stage disperses in barnacles? How is genetic diversity affected? _____

_____

_____

Q15. What animals in the rocky intertidal zone probably reproduce by broadcasting their gametes into the sea? Explain

your choices. _____

_____

_____

Q16. Could a marine animal survive in fresh water? Explain your answer. _____

_____

_____

For humans to extract energy from food by cellular respiration, large pieces of food must first be broken into smaller pieces that can fit in the mouth. These pieces are then chewed into smaller pieces that make their way into the stomach for further digestion.

Q17.   Of the animals examined so far, only anemones, sea stars, and crabs can eat large food. None of these animals has jaws with which to tear apart large food, so how does each of these three animals handle large food to make it suitable for digestion? _____

_____

_____

Habitat 2: Urban Park

Q18.   What would happen to the water in terrestrial animals' bodies if they had no mechanisms for conservation?

_____

_____

_____

Q19.   Why do you think it is important for a planarian to be able to detect and move away from a light source? Could light be dangerous to a flatworm? Explain. _____

_____

_____

Q20.   Because their eyesight is not well developed, what sense must planarians possess to locate prey?

_____

Q21.   Planarians have no organs specialized for respiration or circulation. How do they distribute oxygen and nutrients to their cells? How does this relate to their flattened appearance? _____

_____

_____

Q22.   Would you expect nematodes to have specialized organs for respiration and circulation? Why or why not?

_____

_____

_____

Q23. A parasitic intestinal nematode is bathed in the digestive enzymes of its host. How would you expect the cuticle of this worm to compare with the cuticle of a free-living nematode? _____

_____

_____

Q24. All nematodes have sense organs in their lips, but chemical receptors are found only in free-living nematodes, not in parasitic nematodes. Explain why this difference exists, based on how each type of worm lives. _____

_____

_____

Q25. **Ascaris** infection is typically found in the tropics, but some communities in North America are infected. Based on the method of transmission, what kind of living conditions do infected populations have? How could you avoid infection in such circumstances? _____

_____

_____

Q26. One of the few times earthworms come above ground is when the soil is soaked with water. What do they need above the ground? For what important process is this required? _____

_____

_____

Q27. How are earthworms ill suited for being above ground? Devise two problems earthworms would encounter on the soil's surface. _____

_____

_____

Q28. There are many types of marine annelids. Some are filter feeders; some are active predators. How would each of these types of worms differ from earthworms? Consider mobility, mouthparts, and sensory structures. _____

_____

_____

Ladybird beetles and other insects have small openings in their exoskeleton called **spiracles.** Inside the beetle, the spiracles branch into small tubes that reach all cells of the body.

Q29. What do you suppose these spiracles and branching tubes are for? Why do insects need them? _____

_____

_____

Q30. What could happen to an ectotherm that gets too hot? _____

_____

_____

Q31. How can you tell, just by looking at the jaws, that the squirrel is an herbivore and the snake is a carnivore?

_____

_____

_____

Q32. Why is it impractical for snakes to have fur? Think about how they move and whether they need insulation.

_____

_____

_____

Q33. **Therm** = heat, **endo** = within, **ecto** = outside: explain why endotherms and ectotherms are given these names.

_____

_____

_____

Question

Q34. 85% of all animal species are arthropods. Based on the arthropods you have studied in this lab, as well as your own

experience with arthropods, how would you explain their success? _____

_____

_____

Examine any other live or preserved animal specimens available for you in the lab. By observation and the information you learned in the last exercise, consider how each specimen is adapted for its environment, what kind of locomotion it employs, and how it feeds.

Lab 15   Reproduction/Development
This lab accompanies Chapters 32, 44, and 45 of *Asking About Life.*

Objectives

1. Review life cycle and gamete formation in angiosperms

2. Observe and understand the function of the structures within an angiosperm seed

3. Differentiate between monocots and dicots

4. Understand the formation of gametes in mammals

5. Identify the various stages of early development in a sea star

6. Identify the major organs and structures in early chicken embryos

7. Understand the relatedness between sexually reproducing organisms

Introduction

For both plants and animals, sexual reproduction involves the fusion of a sperm cell with an egg cell to form a single-celled zygote. The process by which this single cell becomes a multicellular organism is called **development.** The genetic information in DNA is used to direct a complex series of events, allowing cell division and differentiation. Cells change their shapes and functions to form specialized organs and other structures. Although each organism follows a different developmental pattern, different organisms have many similarities. In this lab, you will examine gamete formation and early development in angiosperms and animals.

Angiosperms
Materials/Equipment

Dry bean seeds

Soaked bean seeds, at least one per student

Young bean plants (in hypocotyl hook stage with first true leaves attached)

Forceps

Probes

Dissecting microscopes

Plants undergo alternation of generations, in which there is a haploid (1n) gametophyte stage and a diploid (2n) sporophyte stage. As you may recall, the gametophyte is the dominant stage of mosses and their close relatives. In the angiosperms, the sporophyte is the dominant stage; the gametophytes are small structures that exist on the larger sporophyte.

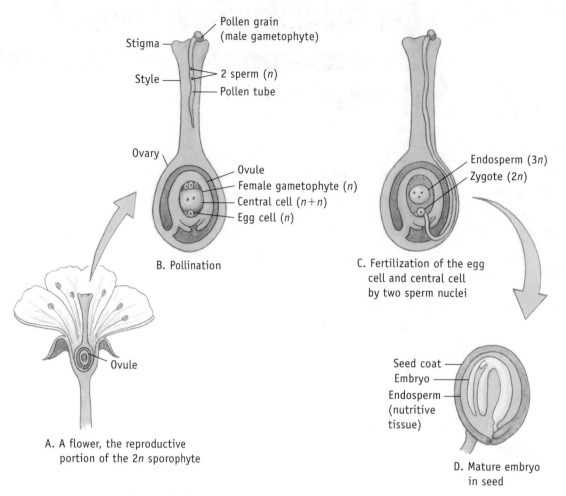

Figure 15-1
From Flower to Seed. A pollen grain lands on the stigma of a flower. A pollen tube grows down the style; two sperm nuclei move down the pollen tube. One sperm nucleus fertilizes the egg cell, creating the zygote that will grow into an embryo. The other sperm nucleus fertilizes the "central cell" in the ovule, producing the endosperm, the nutritive tissue in the seed.

Reproductive Structures

The reproductive structures of angiosperms are located in the flower (see Figure 15-1). Within each ovule in the ovary, meiosis occurs to produce the female gametophyte. Each female gametophyte contains an egg cell (see "How Do Plants Establish a Body Plan?" in the text). Meiosis in the anthers produces the male gametophytes: pollen grains. Pollen is transported to another flower by the wind or an animal pollinator. A sperm cell from the pollen grain fertilizes the egg cell in the ovule. Fertilization results in the formation of a **seed**: an embryo, a food source for the embryo, and a protective seed coat.

> Q1. *If the male and female gametophytes are produced by meiosis, what is their ploidy (1n or 2n)? By what process does a gametophyte produce gametes?*

> Q2. *What is the ploidy of the sperm cell? The egg cell? The embryo?*

At maturity, the seed becomes **dormant**: it dries out, and growth and development are suspended until conditions for growth become favorable.

Germination and Growth of a Dicot

Bean Seed

1. Compare a bean seed that has been soaked in water with a dry bean seed. The first stage of **germination,** or the continuation of growth of a seed, involves taking up water. This process is called **imbibition.**

   Q3. How do the two seeds differ in size and texture?

2. Carefully remove the thin seed coat of the soaked bean seed. You may first need to use a probe to pierce the seed coat. You should now have two white structures joined together. Each of these structures is a **cotyledon.** The cotyledons are the source of nutrients, usually in the form of starch and proteins, for the developing embryo. Angiosperms are divided into two classes based on the number of cotyledons: the **dicots** (beans, carrots, apples, roses, and so on) have two cotyledons in each seed; the **monocots** (tulips, corn, grasses, and so on) have one cotyledon in each seed, as shown in Figure 15-2.

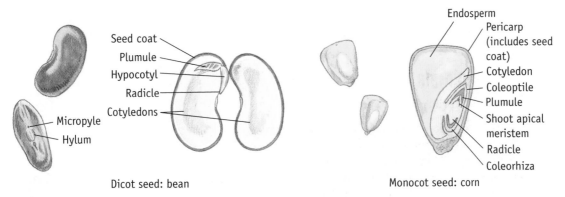

Figure 15-2
Comparison of Typical Dicot and Monocot Seeds. In the bean, the endosperm has been consumed by the developing seed.

   Q4. How do you know the cotyledons contain energy and nutrients? What, primarily, are you eating when you eat beans?

3. Gently separate the two cotyledons. Between the cotyledons is the **embryo.** This tiny structure, given the appropriate conditions, can grow into a bean plant. The embryo will probably remain attached to one of the cotyledons. If you have broken your embryo, get another soaked seed, remove the seed coat, and carefully separate the cotyledons.

4. Place the cotyledon with the embryo on the stage of a dissecting microscope and locate the following regions:

   a. The **plumule** is the embryonic stage of the first true leaves of the plant. It looks like two tiny, folded leaves. Between these two embryonic leaves is the **shoot apical meristem,** which will produce the stems and leaves of the plant.

      Q5. When the plumule leaves emerge from the soil, they will increase in size. What else will they need to develop before they can photosynthesize?

   b. The **hypocotyl** (*hypo* = beneath, *cotyl* = cotyledon) lies just behind the plumule. It will elongate, helping to push the young plant up through the soil.

   c. Beneath the hypocotyl is the **radicle,** the embryonic root. The radicle includes a **root apical meristem,** which will generate the root system of the plant.

5. Draw the embryo and the attached cotyledon, labeling the embryo, the cotyledon, the plumule, the site of the shoot apical meristem, the hypocotyl, the radicle, and the root apical meristem.

Bean Seedlings

1. Examine the young bean seedlings on display. The shoot apical meristem is a small, delicate structure. If it gets damaged as the young plant pushes through the soil, the plant may not be able to grow.

   Q6. What is the first structure to emerge from the soil? How does this protect the shoot apical meristem?

   Q7. What happens to the cotyledons shortly after they emerge from the soil? What must be happening in the cotyledons? Are they still helping to provide the young plant with nutrients?

   Q8. What eventually happens to the cotyledons? Why does the plant no longer need them?

2. Draw one seedling. If there are enough plants, gently remove one from the soil so you can see the root system. Label the leaves, cotyledons, hypocotyl (below the cotyledons), and roots. Indicate from which structures each of these plant parts originated in the seed.

Animals

Materials/Equipment

Microscope slides:
    Mammalian ovary cross section
    Mammalian testis cross section
    Sea star development, showing egg, zygote, different stages of cell division, blastula, and gastrula
    Chicken embryo, 33 hours
    Chicken embryo, 72 hours

Compound microscopes

Lens paper

Chicken eggs:
    One whole egg
    One egg cracked into a clear finger bowl (cover with plastic wrap to prevent it from drying out)

Animal development is divided into eight stages:

1. **Gamete formation:** Production of sperm and eggs by meiosis
2. **Fertilization:** Fusion of an egg and sperm to form a diploid zygote
3. **Cleavage:** Division of the zygote by mitosis into many cells
4. **Germ layer formation:** Folding of embryonic cells to form three germ layers (ectoderm, mesoderm, and endoderm)
5. **Organ formation:** Specialization of cells to form organs such as the heart, kidneys, and nervous system
6. **Growth:** Increase in the size of an organism by mitosis, as well as the formation of materials such as bone, cartilage, and hair
7. **Metamorphosis:** Series of changes during which the organism transforms from a larval to an adult form
8. **Aging:** Further development and eventual death of an organism

In the following exercises, you will explore the first five stages of animal development.

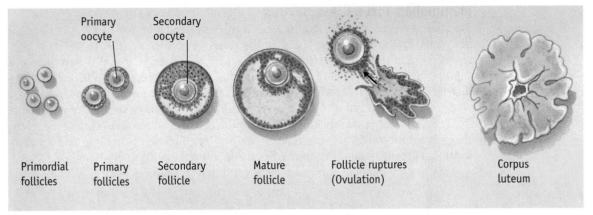

Figure 15-3
Development and Maturation of Primary Oocytes and Follicles. Inside the ovary, the primordial follicles (bottom left) enlarge to form secondary follicles. Inside the follicles, the primary oocytes mature into secondary oocytes. After the follicle ruptures (ovulation), the secondary oocyte migrates into the fallopian tubes, where it may be fertilized by a sperm. The ruptured follicle matures into a corpus luteum.

Mammalian Gonads

In mammals, gametes are formed in structures called **gonads.** Eggs, or **ova** (plural = **ovum**), are formed in the **ovaries,** and sperm are formed in the **testis** (plural = **testes**).

Mammalian Ovary Cross Section

The female gonads, or ovaries, are solid, egg-shaped organs located within the abdominal cavity. A female mammal has two, one on each side. Within the ovary, diploid cells called **primary oocytes** are ready to complete meiosis. A human female is born with all the oocytes she will ever have; at puberty, **follicles** begin to form (see Figure 15-3). Each follicle consists of one primary oocyte surrounded by many **granulosa cells.** Before **ovulation** (release of an ovum), the follicle expands, and the oocyte completes meiosis. The resulting haploid ovum is released when the follicle bursts open. The ovum is ready to be fertilized. The remnant of the follicle after ovulation is called the **corpus luteum.**

Draw and label the structures in bold print.

1. Obtain a prepared slide of a cross section of a mammalian ovary. It will contain follicles in different stages of maturity. Beginning at the lowest power, focus the slide using a compound microscope.

2. The easiest follicles to find are the large, **mature follicles.** The single oocyte is surrounded by several layers of granulosa cells, usually with a space between them. In life, this space was filled with the fluid that would eventually cause the follicle to burst open.

3. The **primary follicles** are smaller, consisting of the central oocyte surrounded by a layer of small granulosa cells.

    Q9. How does the number of mature follicles compare with the number of primary follicles? What accounts for the difference?

4. Locate a **corpus luteum,** the remnant of the follicle after ovulation. This will appear as a large, somewhat solid structure with wavy edges. The corpus luteum produces hormones (progesterone and inhibin) that prepare the female's body for fertilization and pregnancy. If the ovum is not fertilized, the corpus luteum will degenerate.

Mammalian Testis Cross Section

The male gonads, or testes, are located in the **scrotum,** a pouch that lies outside the body to keep the developing sperm at the correct temperature (just below body temperature). Each testis is filled with about 125 meters of tiny coiled **seminiferous tubules.** Along the walls of the seminiferous tubules lie diploid cells called **spermatogonia.** The spermatogonia differentiate into **primary spermatocytes,** which undergo meiosis (see Figure 15-4). Meiosis I results in **secondary spermatocytes;** meiosis II results in **spermatids** that move into the center, or **lumen,** of the tubule. Here, they develop into mature sperm, each with a **flagellum** for swimming and an **acrosome,** an organelle filled with enzymes that allow the sperm to enter an egg.

Starting at puberty, a human male continually produces sperm—about two to four million a day.

Draw and label the structures in bold print.

1. Obtain a prepared slide of a cross section of a mammalian testis. Start on the lowest power of a compound microscope. The circular structures you see are cross sections of **seminiferous tubules.**

2. Work up to the high power lens, and focus on the contents of one tubule. The inner wall is lined with **spermatogonia** that have not yet undergone meiosis; as

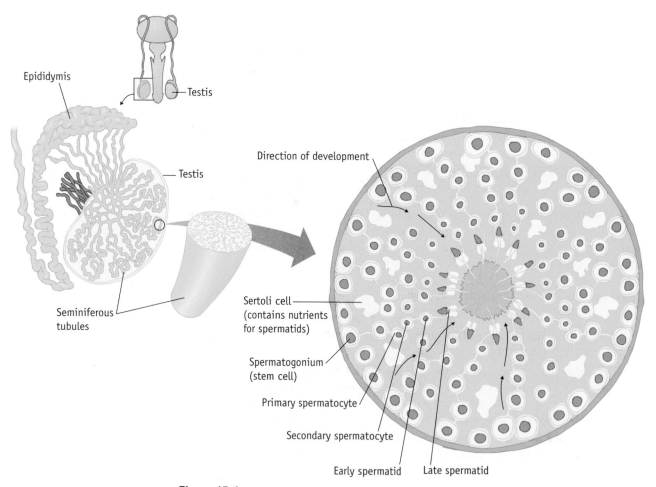

Epididymis

Testis

Testis

Seminiferous tubules

Direction of development

Sertoli cell (contains nutrients for spermatids)

Spermatogonium (stem cell)

Primary spermatocyte

Secondary spermatocyte

Early spermatid

Late spermatid

Figure 15-4
Development and Maturation of Sperm. Inside the testis are hundreds of seminiferous tubules. On the inside wall of each tubule are diploid cells called spermatogonia, which divide, by meiosis, to form four spermatids. For 2-½ months, the spermatids mature into sperm. Sertoli cells supply nutrients to the sperm, and interstitial cells synthesize testosterone. Testosterone increases sex drive and aggressive behavior; it suppresses the secretion of hormones that stimulate the interstitial cells to release testosterone.

you get closer to the center of the tubule, you will see **primary spermatocytes, secondary spermatocytes, spermatids,** and mature **sperm** with flagella. The flagella point toward the lumen of the tubule.

Q10. Are spermatogonia haploid or diploid? What about the spermatids?

Cleavage and Germ Layer Formation

After gamete formation in all animal species, the gametes are brought together in a type of sexual exchange that differs among species. The result of fertilization of an egg by a sperm cell is a single-celled, **diploid zygote.**

The zygote is a large cell with a nucleus no larger than an ordinary cell. Therefore, it has a proportionally large amount of cytoplasm for its plasma membrane and nucleus to regulate. To solve these regulatory problems and to allow cells to specialize, the zygote undergoes a series of rapid divisions in a process called **cleavage,** illustrated in Figure 15-5. The embryo begins to take on the form of a hollow ball, or **blastula.** The structure of the blastula depends on the species.

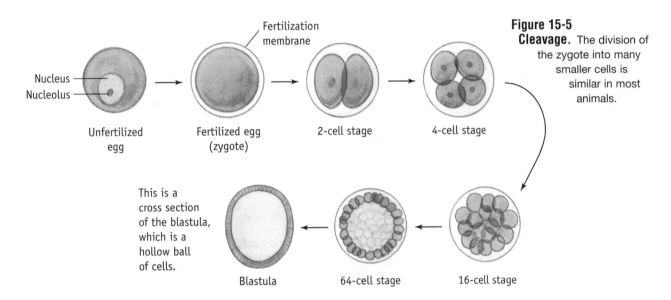

Figure 15-5
Cleavage. The division of the zygote into many smaller cells is similar in most animals.

Next, the embryo forms three **germ layers** (endoderm, mesoderm, and ectoderm) in a process called **gastrulation.** The most visible result of gastrulation is the formation of a "tube within a tube" body plan; the inner tube is the embryonic gut, or **archenteron,** and the embryo is now called a **gastrula.** Imagine pushing your finger into a balloon until it reaches the far wall of the balloon, as shown in Figure 15-6. If you could then poke your finger through that far wall to form another opening, the balloon would look similar to the gastrula. In a real embryo, however, gastrulation involves the migration of individual cells as well as changes in the forms and properties of the cells. The first opening to the archenteron (where you started to poke your finger into the balloon) is called the **blastopore.** If the blastopore becomes the mouth of the adult animal, the animal is a protostome; if the blastopore becomes the anus of the adult animal, the animal is a deuterostome.

Sea Star Development

Cleavage and gastrulation are easily observed in a prepared slide of sea star development. Obtain a slide of sea star development. This is a thick slide; it will break if you try to focus using the high power lens of the compound microscope. **Examine this slide using only the low and medium power objectives.**

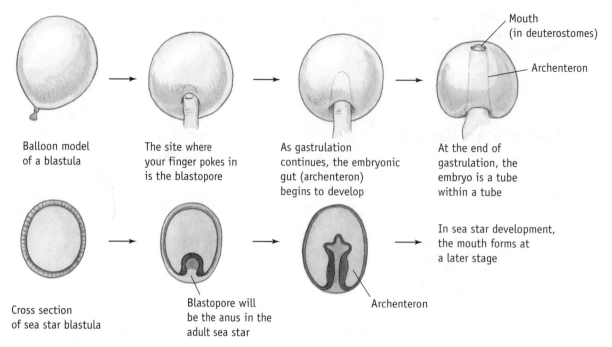

Balloon model
of a blastula

The site where
your finger pokes in
is the blastopore

As gastrulation
continues, the embryonic
gut (archenteron)
begins to develop

At the end of
gastrulation, the
embryo is a tube
within a tube

Mouth
(in deuterostomes)

Archenteron

Cross section
of sea star blastula

Blastopore will
be the anus in the
adult sea star

Archenteron

In sea star development,
the mouth forms at
a later stage

Figure 15-6
Gastrulation. Comparison of gastrulation in a sea star embryo to a model of gastrulation using a balloon.

Locate and draw a representative of each of the following stages, labeling all the structures in bold print that you can.

1. Look for an **unfertilized egg.** It should contain a nucleus and a nucleolus.

2. A fertilized egg, or **zygote,** will have a dense cytoplasm and no obvious nucleus. Adjust the light and focus to see the **fertilization membrane,** which lifts away from the surface of the cell when the sperm penetrates the egg. It appears as an unstained and often wrinkled layer just outside the plasma membrane of the cell.

3. Locate embryos in the process of **cleavage:** find embryos in 2-, 4-, 8-, and 16-cell stages.

   Q11. Is the fertilization envelope still intact during these stages?

   Q12. Are these cells diploid or haploid?

4. As cleavage continues, the cells can no longer adhere to each other, and a hollow space will form in the center of the ball of cells. The resulting hollow ball of cells is a **blastula.** If you focus on a blastula and move the fine focus knob slowly up and down, you can focus through the blastula to see its surface as well as its hollow interior.

   Q13. How does the size of the blastula compare with the size of the zygote?

   Q14. How does the size of the cells that comprise the blastula compare with the size of the zygote? Has any growth occurred?

   Q15. Is there any advantage to having many smaller cells rather than a single large one? If so, what are two advantages?

5. Find embryos in the process of **gastrulation.** Notice how it begins with the formation of a **blastopore** as cells begin to migrate into the hollow center of the embryo. As gastrulation continues, the **archenteron** continues to extend across the interior of the embryo. When it reaches the opposite side, the tube within a tube is complete, and there are three germ layers. The **ectoderm** will form the outer skin and the nervous system; the **mesoderm** will give rise to the muscles, connective tissues, and reproductive organs; the **endoderm** will form the diges-

tive tract and associated organs. The mesoderm develops last and is best seen in a mature gastrula between the other two germ layers.

Q16. *What germ layer lines the archenteron?*

Q17. *What is the outermost germ layer on the embryo?*

Q18. *Sea stars are in the phylum Echinodermata, and echinoderms are deuterostomes. Will the blastopore become the mouth or anus of this animal?*

Q19. *Does the embryo grow during gastrulation?*

6. Because sea stars sexually reproduce by broadcasting their gametes into the ocean, the developing embryo gets no care or nourishment from its mother. In addition, the egg contains very little yolk. After gastrulation, the embryo develops into a microscopic larval stage that can swim and obtain food (small particles of organic matter it filters from the seawater). See Figure 15-7. It uses this energy to complete its growth and development until it **metamorphoses** into an adult. There may be larval sea stars on your slide.

Q20. *Does the larva look like a sea star? Does it feed like an adult sea star (opening and digesting shellfish and other invertebrates)? If your answer is no, why can't it feed like an adult?*

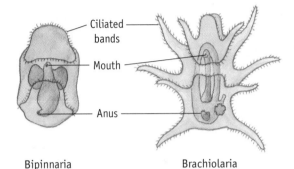

Bipinnaria   Brachiolaria

Figure 15-7 Sea Star Development. A sea star gastrula develops into a bipinnaria, a larval stage that uses ciliated bands for swimming in the ocean. The bipinnaria changes into a brachiolaria with arms for attachment to a substrate when the larva settles and metamorphoses into an adult.

Organ Formation (Organogenesis)

Once an animal embryo has established three germ layers, it begins to form organs. The chicken is a good example of early development in vertebrates. The early development in a vertebrate is slightly different from that in the sea star.

1. **Chicken egg**
   a. Examine the cracked chicken egg on display. What looks like the yolk is actually the egg cell, or ovum. The yolk (mostly protein and fat) is the food reserve in the cytoplasm of the ovum. On each side of the ovum is a white **chalaza,** a small cord made of protein fibers that holds the yolk in place by anchoring it to either end of the egg.
   b. Look for a small white spot on the surface of the yolk. If the egg is unfertilized (like most commercial eggs), this spot is the nucleus of the egg cell.

   Q21. *How many cells are in the unfertilized egg?*

2. When a chicken egg is fertilized, the embryo develops on the surface of the large yolk. The presence of that yolk greatly influences the way early development occurs. Early cleavage results in a disc of cells, called a **blastodisc,** instead of a blastula. Gastrulation occurs when cells move into a groove (the **primitive streak**), equivalent to the blastopore of the sea star embryo. See Figure 15-8.

   Q22. *Birds and reptiles both produce eggs with large yolks; most mammals, which have small eggs with few food reserves, follow patterns of gastrulation and organogenesis similar to those of birds and reptiles. How would you account for this similarity?*

3. After the three germ layers are formed, the edges of the disk-shaped embryo fold downward and come together to form the embryo into a three-layered tube. The endoderm, mesoderm, and ectoderm layers then undergo a sequence of folding, cell movement, and change in cell shape to form the organs.

   All vertebrates share certain features that place them into the phylum Chordata. These four chordate characteristics are present at some point in

Figure 15-8
Early Cleavage and Gastrulation in a Chicken Embryo.

These stages take place on the surface of the yolk. Note the similarities and differences between these developmental stages in a chicken and in a sea star.

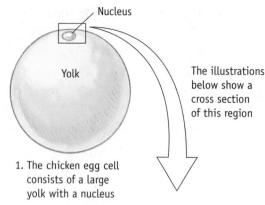

Nucleus

Yolk

The illustrations below show a cross section of this region

1. The chicken egg cell consists of a large yolk with a nucleus

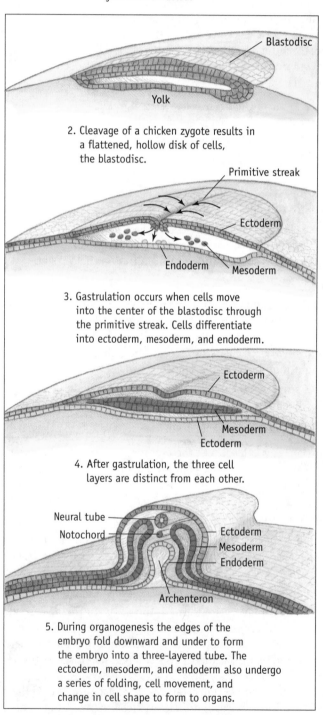

Blastodisc

Yolk

2. Cleavage of a chicken zygote results in a flattened, hollow disk of cells, the blastodisc.

Primitive streak

Ectoderm

Endoderm

Mesoderm

3. Gastrulation occurs when cells move into the center of the blastodisc through the primitive streak. Cells differentiate into ectoderm, mesoderm, and endoderm.

Ectoderm

Mesoderm

Ectoderm

4. After gastrulation, the three cell layers are distinct from each other.

Neural tube

Notochord

Ectoderm

Mesoderm

Endoderm

Archenteron

5. During organogenesis the edges of the embryo fold downward and under to form the embryo into a three-layered tube. The ectoderm, mesoderm, and endoderm also undergo a series of folding, cell movement, and change in cell shape to form to organs.

Tunicate larva (urochordate)

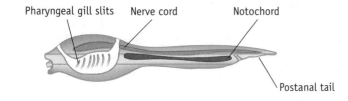

Amphioxus (cephalochordate)

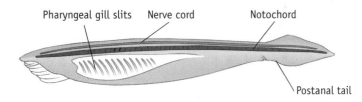

Lizard (vertebrate)

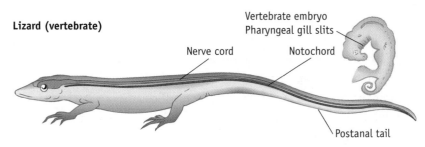

Figure 15-9
Four Features of Chordates. The three chordate subphyla—the tunicates (sea squirts), the cephalochordates (amphioxus), and the vertebrates—all share four distinguishing features: pharyngeal gill slits, a dorsal nerve cord, a notochord, and a postanal tail.

each vertebrate's life cycle, whether or not they persist into adulthood. See Figure 15-9.

- **Notochord:** A flexible cartilaginous rod running down the dorsal surface provides back support for invertebrates and some vertebrates. In most vertebrates, it is replaced by a bony or cartilaginous backbone.
- **Dorsal hollow nerve cord:** This cord runs between the notochord and the back. It becomes the spinal cord and brain in the adult.
- **Pharyngeal slits, or gill slits:** Holes in the sides of the body are used for filter feeding, respiration, or other functions, depending on the species.
- **Postanal tail:** The tail extends beyond the anus.

In the following two stages of chicken embryonic development, you will be able to see most of these characteristics.

4. **33-hour chicken embryo:** Obtain a prepared slide of a 33-hour chicken embryo (33 hours refers to the number of hours after fertilization). See Figure 15-10. At this stage, the development of the anterior end of the embryo is somewhat ahead of the development of the posterior of the embryo. This slide is very thick and must be viewed at low and medium power only to avoid damaging the high power lens of the microscope.

Draw your specimen, labeling all structures in bold print that you can.
a. At the anterior end of the embryo, locate the developing **brain,** two lobes of which form the **optic vesicles.** The optic vesicles will cause the formation of the eyes. Box 45-1 in the text illustrates this process.

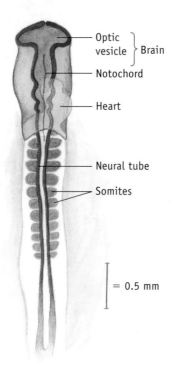

— Optic
vesicle } Brain

— Notochord

— Heart

— Neural tube

— Somites

$\rceil$ = 0.5 mm

**Figure 15-10
33-Hour Chicken Embryo,
with Major Structures
Labeled.**

b. Posterior to the brain is the developing **heart,** which is difficult to distinguish at this stage.

c. Behind the heart, identify the **neural tube,** which will become the entire nervous system. In the central portion of the embryo, the neural tube will become the **dorsal hollow nerve cord.** Just below (ventral to) the neural tube is the **notochord,** which appears as a dark thread under the brain in this preparation.

   At this stage, the neural tube is still in the process of folding into a tube shape: at its anterior end, it has folded into a hollow tube; at the posterior end, the folds have not yet met at the center.

d. Along the neural tube are several paired structures. These are **somites,** formed from mesoderm, which will generate bone and muscle groups.

e. At the posterior end of the embryo, the remains of the **primitive streak** may be seen.

5. **72-hour chicken embryo:** Obtain a prepared slide of a 72-hour chicken embryo. See Figure 15-11. Also view this slide at low and medium power only. The embryo has turned almost completely onto its left side. Its head and upper body are curving, giving the embryo the shape of a backward question mark.

Draw your specimen, labeling all the structures in bold print that you can.

a. The **brain** is more distinct and divided into several regions.

b. The **eye** now has distinctive parts: a lens and an optic cup.

c. Locate the **otic vesicle,** which will become the ear.

d. In the center of the embryo is the **heart,** which is growing quickly and actively pumping blood at this stage.

e. You may be able to see the **pharyngeal slits** between the heart and the otic vesicles. In a fish, these slits would become the gill slits; in terrestrial vertebrates, they form other structures. For example, the bones in the middle ear of mammals are formed from the pharyngeal slits.

f. Look for lobes of tissue that are the **limb and tail buds:** these are the portions of the embryo that will develop into its legs, wings, and tail. The tail bud marks the end of the embryo; the leg buds will lie just above this; the wing buds are closer to the center of the embryo.

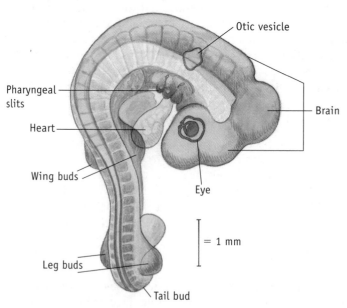

Otic vesicle

Pharyngeal
slits

Heart

Wing buds

Brain

Eye

= 1 mm

Leg buds

Tail bud

**Figure 15-11
72-Hour Chicken Embryo, Showing Major Structures.**

Reproduction/Development
Data Sheet

Angiosperms
Reproductive Structures

Q1.   *If the male and female gametophytes are produced by meiosis, what is their ploidy (1n or 2n)? By what process does a*

   gametophyte produce gametes? _____

   _____

   _____

Q2.   *What is the ploidy of the sperm cell? The egg cell? The embryo?* _____

   _____

   _____

Bean Seed

Q3.   *How do the bean seed that has been soaked in water and the dry bean seed differ in size and texture?* _____

   _____

   _____

Q4.   *How do you know the cotyledons contain energy and nutrients? What, primarily, are you eating when you eat beans?*

   _____

   _____

   _____

Q5.   *When the plumule leaves emerge from the soil, they will increase in size. What else will they need to develop before*

   they can photosynthesize? _____

   _____

   _____

Draw the embryo and the attached cotyledon, labeling the embryo, cotyledon, the plumule, the site of the shoot apical meristem, the hypocotyl, the radicle, and the root apical meristem.

Bean Seedlings

Q6.   *What is the first structure to emerge from the soil? How does this protect the shoot apical meristem?* _____

_____

_____

Q7.   *What happens to the cotyledons shortly after they emerge from the soil? What must be happening in the cotyledons?*

Are they still helping to provide the young plant with nutrients? _____

_____

_____

Q8.   *What eventually happens to the cotyledons? Why does the plant no longer need them?* _____

_____

_____

Draw one seedling. Label the leaves, cotyledons, hypocotyl (below the cotyledons), and roots. Indicate from which structures each of these plant parts originated in the seed.

Animals
Mammalian Ovary Cross Section

Draw and label the following structures: mature follicles, primary follicles, and corpus luteum.

Q9. *How does the number of mature follicles compare with the number of primary follicles? What accounts for the*

difference? _____

_____

_____

Mammalian Testis Cross Section

Draw and label the following structures: seminiferous tubules, spermatogonia, primary spermatocytes, secondary spermatocytes, spermatids, and sperm.

Q10. *Are spermatogonia haploid or diploid? What about the spermatids?* _____

_____

_____

Sea Star Development

Locate and draw representatives of each stage, labeling all the structures you can.

Q11. Is the fertilization envelope still intact during these stages? _____

Q12. Are these cells diploid or haploid? _____

Q13. How does the size of the blastula compare with the size of the zygote? _____

_____

_____

Q14. How does the size of the cells that comprise the blastula compare with the size of the zygote? Has any growth

occurred? _____

_____

_____

Q15. Is there any advantage to having many, smaller cells rather than a single large one? If so, what are two advantages?

_____

_____

_____

Q16. What germ layer lines the archenteron? _____

Q17. What is the outermost germ layer on the embryo? _____

_____

_____

Q18. Sea stars are in the phylum Echinodermata, and echinoderms are deuterostomes. Will the blastopore become the

mouth or anus of this animal? _____

Q19. Does the embryo grow during gastrulation? _____

Q20. Does the larva look like a sea star? Does it feed like an adult sea star (opening and digesting shellfish and other in-

vertebrates)? If your answer is no, why can't it feed like an adult? _____

_____

_____

Chicken Egg

Q21. How many cells are in the unfertilized chicken egg? _____

Q22. *Birds and reptiles both produce eggs with large yolks; most mammals, which have small eggs with few food reserves, follow patterns of gastrulation and organogenesis similar to those of birds and reptiles. How would you account for this similarity?* _____

_____

_____

33-Hour Chicken Embryo

Draw your specimen, labeling all the structures you can.

72-Hour Chicken Embryo

Draw your specimen, labeling all the structures you can.

Questions

Q23.   A sea star develops from a zygote to a swimming larva in a matter of hours; a chicken takes about 20 days from

fertilization to hatching. Why must the sea star hatch so quickly? _____

_____

_____

Q24.   Why is a larval stage unnecessary for chickens? _____

_____

_____

Q25.   Why do human embryos not need the energy from a large yolk for development? _____

_____

_____

Q26.   Many animals eat seeds, generally a good source of nutrients and calories. Explain why seeds are packed with energy.

_____

_____

_____

Q27.   The angiosperm seed and the shelled egg are both adaptations to life on land. What are three ways in which these

structures are suited to life on land? _____

_____

_____

Q28.   What are the functions of the seed coat and the cotyledons? What structures provide similar functions in a chicken

egg? _____

_____

_____

Q29.   Do you think a human embryo has a tail at some stage of development? How do you know, based on information

from this lab? _____

_____

_____

Appendix

Scientific Notation

Scientific notation is a system used to easily denote numbers that are very large or very small. For instance, 1,000,000 is expressed in scientific notation as 1×10^6, where 6 (the exponent) signifies how many decimal places to the **right** the decimal point is from the 1. Here are some other examples:

$$1.5 \times 10^2 = 150$$

$$3.5 \times 10^5 = 350,000$$

$$7.32 \times 10^{10} = 73,200,000,000$$

If the exponent is a negative number, then it denotes the number of decimal places to the **left** the decimal point would be moved. For example:

$$1.5 \times 10^{-2} = 0.015$$

$$6.43 \times 10^{-6} = 0.00000643$$

$$2.29 \times 10^{-9} = 0.00000000229$$

To multiply two numbers in scientific notation, multiply the numbers and add the exponents. For example:

$$(2.5 \times 10^2)(2 \times 10^4) = 5 \times 10^6$$

$$(3 \times 10^4)(2 \times 10^{-2}) = 6 \times 10^2$$

$$(1.5 \times 10^{-4})(2 \times 10^{-6}) = 3 \times 10^{-10}$$

To divide two numbers in scientific notation, divide the numbers and subtract the exponent of the numerator from the exponent of the denominator:

$$\frac{6 \times 10^4}{3 \times 10^2} = 2 \times 10^2$$

$$\frac{3 \times 10^3}{2 \times 10^5} = 1.5 \times 10^{-2}$$

$$\frac{9.3 \times 10^6}{3 \times 10^{-2}} = 3.1 \times 10^8$$

$$\frac{7 \times 10^{-3}}{3.5 \times 10^{-5}} = 2 \times 10^2$$

Perform the following computations:

1. Write these numbers in decimals:
   a. $1 \times 10^4 =$
   b. $3.5 \times 10^{-6} =$
   c. $2.7 \times 10^5 =$
   d. $6.74 \times 10^{-3} =$

2. Write these numbers in scientific notation:
   a. $0.002 =$
   b. $33,760,000 =$
   c. $0.00000045 =$
   d. $150 =$
   e. $85,200,000,000 =$

3. Multiply the following numbers:
   a. $(3 \times 10^2)(2 \times 10^4) =$
   b. $(4.2 \times 10^3)(3 \times 10^2) =$
   c. $(9 \times 10^5)(1 \times 10^{-3}) =$
   d. $(5 \times 10^{-4})(2.5 \times 10^{-5}) =$

4. Divide the following numbers:

   a. $\dfrac{5 \times 10^6}{2 \times 10^2} =$

   b. $\dfrac{6.6 \times 10^4}{3 \times 10^5} =$

   c. $\dfrac{8.4 \times 10^3}{4.2 \times 10^{-6}} =$

   d. $\dfrac{4 \times 10^{-4}}{5 \times 10^{-2}} =$

Metric System

In the United States, we use the old British system of weights and measures that carried over from colonial times. Units of **length** are as follows: there are 12 inches in a foot, 3 feet in a yard, 5.5 yards in a rod, 40 rods in a furlong, and 8 furlongs in a mile. The inch is divided into 16ths and sometimes into 32nds. Conversions among these units are cumbersome. For instance, if you want to determine how many inches are in a mile, you have to do this calculation:

$$1 \text{ mile} = (8 \text{ furlongs/mile}) \times (40 \text{ rods/furlong}) \times (5.5 \text{ yards/rod}) \times (3 \text{ feet/yard})$$
$$\times (12 \text{ inches/foot}) = 63,360 \text{ inches/mile}$$

Similarly, the units of **weight** are 27.34375 grains in a dram, 16 drams in an ounce, 16 ounces in a pound, and 2,000 pounds in a short ton (2,240 pounds in a long ton). Conversions among these units also require some effort.

The units for measuring **volume** are as follows: 8 fluid ounces in a cup, 2 cups in a pint, 2 pints in a quart, and 4 quarts in a gallon. How many fluid ounces in a gallon? Before calculating, read on.

Most other countries, and scientists all over the world, use the **metric system,** an international decimal system of measurement. The basic unit of length is the **meter,** the basic unit of mass (or weight) is the **gram,** and the basic unit of volume is the **liter.**

In the metric system, the divisions of units are not arbitrary (like 12 inches in a foot and 3 feet in a yard). Instead, everything is based on powers of 10, so conversions among metric units involve moving the decimal point, rather than complicated calculations.

The prefixes (and their abbreviations) for the divisions and multiples of the basic metric units are as follows:

Giga- (G)	1,000,000,000 (1×10^9)
Mega- (M)	1,000,000 (1×10^6)
Kilo- (k)	1,000 (1×10^3)
Hecto- (h)	100 (1×10^2)
Deca- (D)	10
Deci- (d)	0.1 (1×10^{-1})
Centi- (c)	0.01 (1×10^{-2})
Milli- (m)	0.001 (1×10^{-3})
Micro- (μ)	0.000001 (1×10^{-6})
Nano- (n)	0.000000001 (1×10^{-9})

This system was first introduced by the Paris Academy of Sciences in 1791 and was received in the United States in 1893. In 1975, President Ford signed the Metric Conversion Act. Although most cars, measuring cups, and food products carry metric units, the prominent units on the labels are still those of the British system.

Conversions between Metric Units

To convert units, write all your steps and include units in every step. Because the metric system is a decimal system, conversion between units involves changing the power of 10 (or the location of the decimal point).

The most common metric prefixes used in this course will be centi-, milli-, and micro-. For instance, you will measure volume in milliliters; length in centimeters, millimeters, or micrometers; and weight in grams and milligrams. Because you know there are 100 cm in 1 m, 1,000 mm in 1 m, and 10 mm in 1 cm, conversions among these units are simple. For instance, if you measure a caterpillar that is 4.5 cm and want to know how many millimeters it is, just multiply 4.5 by 10:

$$4.5 \text{ cm} \times \frac{10 \text{ mm}}{1 \text{ cm}} = 45 \text{ mm}$$

Note that the centimeters cancel each other out, leaving the units in millimeters.

If you do not know these conversions and want to convert between centimeters and meters, you must make a conversion factor based on the information in the prefix list. It states that 1 cm = 1×10^{-2} m. This equation can be used as a conversion factor by writing it as a fraction, with either side as the numerator:

$$\frac{1 \times 10^{-2} \text{ m}}{1 \text{ cm}} = \frac{1 \text{ cm}}{1 \times 10^{-2} \text{ m}}$$

To convert between meters and centimeters, use the conversion factor that will give the units you want. If you want to know how many centimeters are in 2.3 m, multiply 2.3 m by the conversion factor in which the meters will cancel each other out:

$$2.3 \text{ m} \times \frac{1 \text{ cm}}{1 \times 10^{-2} \text{ m}} = 2.3 \times 10^2 \text{ cm} = 2,300 \text{ cm}$$

Conversely, to find out how many meters are in 45 cm:

$$45 \text{ cm} \times \frac{1 \times 10^{-2} \text{ m}}{1 \text{ cm}} = 45 \times 10^{-2} \text{ m} = 0.45 \text{ m}$$

Similarly, to convert between other metric units, you must determine a conversion factor. If you want to convert 3 mm into nanometers, you know from the list that:

$$1 \text{ mm} = 1 \times 10^{-3} \text{ m}$$

$$1 \text{ nm} = 1 \times 10^{-9} \text{ m}$$

Use these equations to make a conversion factor that will give you the number of millimeters in a nanometer:

$$\frac{1 \times 10^{-3} \text{ m}}{1 \text{ mm}} \times \frac{1 \text{ nm}}{1 \times 10^{-9} \text{ m}} = \frac{1 \times 10^{6} \text{ nm}}{1 \text{ mm}}$$

Now that you have a conversion factor, multiply it by the 3 mm:

$$3 \text{ mm} = \underline{\hspace{2cm}} \text{ nm?}$$

$$3 \text{ mm} \times \frac{1 \times 10^{6} \text{ nm}}{1 \text{ mm}} = 3 \times 10^{6} \text{ nm}$$

Do the following conversions:

37.6 μm = \underline{\hspace{2cm}} cm

56 L = \underline{\hspace{2cm}} mL

23.14 g = \underline{\hspace{2cm}} kg

Conversions from Standard to Metric

When you do these conversions, write down all your steps and make sure to include units. This makes it easier to track what you are doing.

For example, if you are asked to convert 12 miles into meters, given the conversion factor of 1.6 km in a mile, write down what you know:

12 miles = how many meters?
1.6 km/mile
1,000 m/km

First, take the 12 miles and determine how many kilometers this is, because you know this conversion:

$$12 \text{ miles} \times \frac{1.6 \text{ km}}{1 \text{ mile}} = 19.4 \text{ km}$$

Notice that the miles cancel each other out, and you are left with the units of kilometers, which you want.

The question asked for the number of **meters** in 12 miles, and you know that there are 1×10^3 m in 1 km, so:

$$(19.4 \text{ km}) \times \frac{(1 \times 10^3 \text{ m})}{1 \text{ km}} = 19.4 \times 10^3 \text{ m} = 19,400 \text{ m}$$

1. If there are 2.54 cm in an inch, what is your height in centimeters? How many meters is this? Note: You must first convert your height into inches. There are 12 inches in a foot.

2. There are 3,790 mL in a gallon. How many liters of milk are in a gallon?

3. There are 454 g in a pound. What is the weight of a 120-pound woman in kilograms?

Measuring Temperature

In the British system, temperature is measured in degrees Fahrenheit (°F). The metric system uses degrees Celsius (°C). Conversions between the two systems are made using the following formulas:

$$°C = \frac{5}{9}(x°F - 32) \qquad\qquad °F = \frac{9}{5}(x°C) + 32$$

Perform the following conversions:

1. Human body temperature is 98.6°F. What is this temperature in Celsius?
2. Water boils at 212°F. At what temperature does water boil in Celsius?
3. Water freezes at 0°C. What is the freezing point of water in Fahrenheit?
4. A comfortably warm day is 22.2°C. What is this temperature in Fahrenheit?

Conversion Table: Metric/British

1 inch = 2.54 cm	1 cm = 0.394 inches
1 foot = 0.305 m	1 m = 3.28 feet
1 mile = 1.61 km	1 km = 0.621 miles
1 fluid ounce = 29.6 mL	1 mL = 0.0338 fluid ounces
1 pint = 0.474 L	1 L = 2.11 pints
1 gallon = 3.79 L	1 L = 0.264 gallons
1 ounce = 28.35 g	1 g = 0.0353 ounces
1 pound = 0.454 kg	1 kg = 2.2 pounds

$$°C = \frac{5}{9}(x°F - 32) \qquad °F = \frac{9}{5}(x°C) + 32$$

Dependent/Independent Variables

The **variables** in an experiment are the conditions expected to change during the experiment. If you want to see how the amount of fertilizer affects the growth of bean plants, you would grow several bean plants. In the control group, you would add no fertilizer; in the experimental groups, you would add different amounts of fertilizer. Over time, you would measure an indicator of growth (for example, the height of the plants). The **independent variable** is the scientific condition manipulated in an experiment; in this case, it would be the amount of fertilizer. The **dependent variable** is the scientific condition measured in response to changes in the independent variable. In the bean plant experiment, the dependent variable is the height of the plants.

An experiment can have more than one dependent variable. Healthy plant growth may be measured not only by height but also by the number of leaves produced, the weight of the plant, the number of bean pods produced, and the number of seeds per pod. As the designer of the experiment, you can choose to measure one or more of these variables as measures of growth.

Conversely, an experiment often has only one independent variable. In the bean experiment, many conditions could affect plant growth: the amount of light, the amount of water, and temperature, to name three. To know which variable is affecting the dependent variable, limit the independent variables. If you grow all plants in the same conditions except for the amount of fertilizer, you can be fairly certain the fertilizer is the cause of differences in growth.

Identify the dependent and independent variables in the following experiments:

1. Mice are fed the same amount of four different brands of feed for three weeks. Their weight is taken at the beginning and the end of the three weeks.

2. Oxygen production by a plant is measured in the dark, 1 m from a light source and 0.5 m from a light source.

3. Sets of grass seeds are exposed to different degrees of heat. They are then planted. The seeds that germinate are counted.

4. Researchers count the different species of animals and algae that settled and grew on various surfaces placed in the ocean.

Graphs

Sometimes it is useful to present the results of an experiment in the form of a graph. A graph is a visual representation of the results; it shows relationships between an independent variable and a dependent variable. Specifically, it shows how the independent variable affects the dependent variable. The information should be presented in a way that is easy to interpret.

A graph should be drawn on graph paper or a grid, following these rules:

1. Give the graph a title that describes the experiment.

2. The independent variable is usually plotted on the **x-axis** (horizontal axis); the dependent variable is usually plotted on the **y-axis** (vertical axis), as shown in Figure A-1.

3. Title each axis with the variable and the units you used to measure that variable.

4. Label the intervals on each axis with numbers that are evenly spaced and correspond with your data. For instance, if the height of your bean plants in the fertilizer experiment ranged from 12 cm to 37 cm, you might choose a vertical axis that goes from 0 to 40 cm in intervals of 5 cm. Make sure that most of the space on your graph is used. If the largest value on the height axis is 37 cm, do not make a height axis that goes to 100 cm: over half of your graph would be unused.

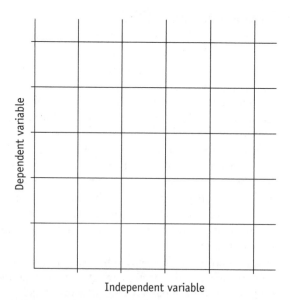

Figure A-1
Graph. A graph usually has the independent variable plotted along the horizontal axis and the dependent variable plotted along the vertical axis.

Line Graphs and Bar Graphs

The most common types of graphs are **bar graphs** and **line graphs**. Line graphs are used for continuous data, such as changes in growth over time. Each data point is plotted on the graph, and the points are connected with lines, as shown in Figure A-2.

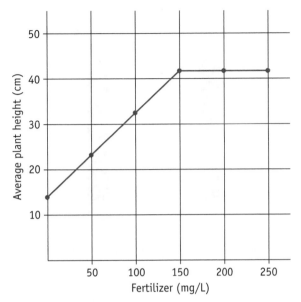

Figure A-2
Line Graph for Fertilizer Experiment. The height of bean plants grown using different concentrations of fertilizer.

Two sets of data may be shown on the same graph by using different types of markers to indicate the different data points (see Figure A-3). Always include a key when making a line graph with more than one set of data.

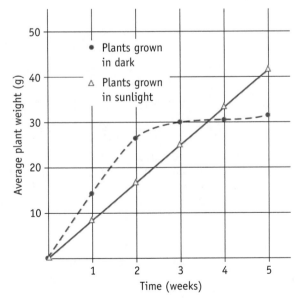

Figure A-3
Line Graph with Two Sets of Data. The rate of plant growth by weight for bean plants grown in the light and for bean plants grown in the dark.

A bar graph is used for discrete data in which there are no intermediate values. In this experiment, goldfish were raised in pond water or tap water. The weights of males and females in each group of fish were recorded after three months, as shown in Figure A-4.

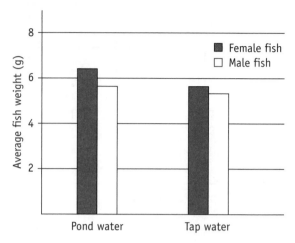

Figure A-4
Bar Graph. The average weights of male and female goldfish raised in pond water and in tap water.

These are discrete sets of data: the independent variable is one of two water types with no intermediates; the goldfish are either male or female.

Interpreting Graphs

Figure A-5 shows three possibilities of results for the bean plant and fertilizer experiment. Note that the units are the same for each graph. Before reading on, consider the type of results each graph shows.

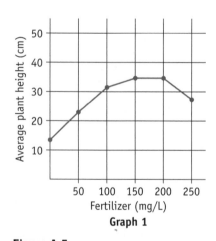
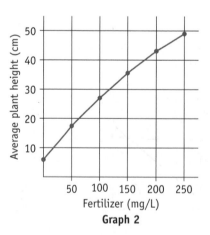
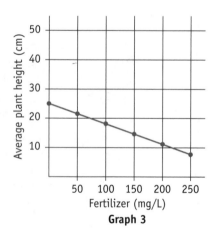

Figure A-5
Three Graphs Showing Different Results. The height of bean plants grown using different concentrations of fertilizer.

In Graph 1, the line rises, levels off, and then drops. This shows that an increasing amount of fertilizer improved plant growth to a certain point. After that point, the effect of increasing fertilizer was insignificant until the fertilizer reached a concentration that actually inhibited plant growth.

Q1. At what fertilizer concentration did plant growth stop improving?

Q2. What was the height of the tallest plant grown in this experiment?

In Graph 2, the line rises dramatically. This shows that for the concentrations of fertilizer used, increasing amounts of fertilizer dramatically increased plant growth.

Q3. *What was the height of the smallest plant grown in this experiment?*

In Graph 3, the line slopes downward. In these results, increasing amounts of fertilizer inhibited plant growth.

Draw the line on the graph in Figure A-6 for the results of this experiment if all plants grew to about 22 cm (the amount of fertilizer added made no difference).

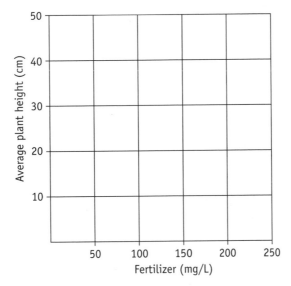

Figure A-6
Height of Bean Plants Grown Using Different Concentrations of Fertilizer.

When you are interpreting a graph, always consider the units and intervals used to label the axes. Are they appropriate? Consider the graphs in Figure A-7.

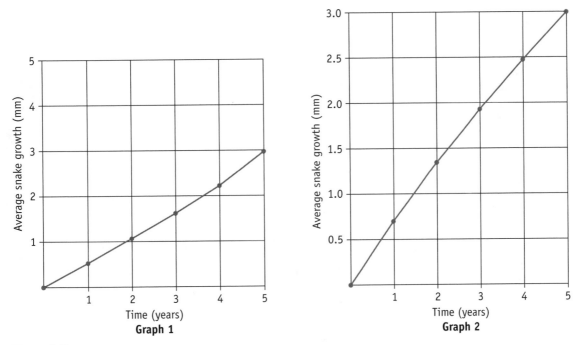

Figure A-7
Two Representations of the Same Data. Snake growth in millimeters over five years.

Notice that although these graphs show exactly the same data, the way the intervals were chosen makes them look very different. If you were not paying attention and looked only at Graph 2, you might think the snake grew quite a bit over the years. But the snake only grew 3 mm over five years; the intervals chosen exaggerated the way the amount of growth appears on the graph. When you are presented with graphs in the media, always note the intervals and units used to see if the data are presented fairly (imagine that Graph 2 presented gasoline prices or unemployment rates).

References

Benton, William, publisher 1966. *Encyclopaedia Britannica,* vols. 15 and 23, Encyclopaedia Britannica, Inc.

Campbell, Neil A. 1996. *Biology,* 4th edition, Benjamin/Cummings Publishing Company, Inc.

Dickey, Jean. 1995. *Laboratory Investigations for Biology,* Benjamin/Cummings Publishing Company, Inc.

Evert, Ray F., and Eichhorn, Susan E. 1992. *Laboratory Topics in Botany,* to accompany Raven, Evert, and Eichhorn, *Biology of Plants,* 5th edition, Worth Publishers, Inc. (Onion DNA extraction protocol was modified from this protocol.)

Gleason, Henry A. 1952. *New Britton and Brown Illustrated Flora,* vols. 1 and 2, Lancaster Press, Inc.

Gleason, Henry A. 1952. *Illustrated Flora of the Northeastern United States and Adjacent Canada,* Lancaster Press, Inc.

Hickman, C.P. Sr., Hickman, C.P. Jr., and Hickman, F.M. 1970. *Integrated Principles of Zoology,* 5th edition, C.V. Mosby Company.

Linnan, M.J., et al. 1988 September 29. *New England Journal of Medicine* 319(13)823-8.

Mader, Sylvia S. 1998. *Biology Laboratory Manual,* 6th edition, WCB/McGraw-Hill. (Yeast fermentation protocol was modified from this protocol.)

Morgan, Judith G., and Carter, M. Eloise B. 1996. *Annotated Instructor's Edition for Investigating Biology,* 2nd edition, Benjamin/Cummings Publishing Company, Inc.

Niesen, Thomas M. 1982. *The Marine Biology Coloring Book,* Harper Perennial.

Pearse, D.K., et al. 1987. *Living Invertebrates,* Boxwood Press.

Peters, David. 1991. *From the Beginning: The Story of Human Evolution,* Morrow Junior Books.

Raven, Peter H., Evert, Ray F., and Eichhorn, Susan E. 1992. *Biology of Plants,* 5th edition, Worth Publishers, Inc.

Strathmann, M.E. 1987. *Reproduction and Development of Marine Invertebrates of the Northern Pacific Coast,* University of Washington Press.

Web Sites

www.accessexcellence.org (Accessed December 1999.)

Backyardnature.net/squirrels.html (Accessed June 2003.)

www.biosci.ohio-state.edu/~parasite/ascaris.html (Accessed June 2003.)

cas.bellarmine.edu/tietjen/images/nematodes.htm (Accessed June 2003.)

www.cdc.gov (Accessed August 2003.)

Chickscope.beckman.uinc.edu/explore/embryology/ (Accessed January 2000.)

www.comeunity.com/adoption/health/parasites/parasites-NIH.html (Accessed June 2003.)

dbhs.wvusd.k12ca.us/sigfigRules.html (Accessed February 2000.)

www.exploratorium.edu (Accessed January 2003.)

www.foodsafety.gov (Accessed January 2003.)

Foodsafety.org (Accessed January 2003.)

martin.parasitology.mcgill.ca/jimspage/ascaris.htm (Accessed July 2003.)

www.nlm.nih.gov (Accessed July 2003.)

Orni.gov (Accessed July 2003.)

www.squirrels.org/gray.html (Accessed July 2003.)

Photo Credits

This page constitutes an extension of the copyright page. We have made every effort to trace the ownership of all copyrighted material and to secure permission from copyright holders. In the event of any question arising as to the use of any material, we will be pleased to make the necessary corrections in future printings. Thanks are due to the following authors, publishers, and agents for permission to use the material indicated.

Chapter 1
3: Emery Photography, Inc., Columbus, Ohio. Courtesy of Parco Scientific

Chapter 3
28: Emery Photography, Inc., Columbus, Ohio. Courtesy of Parco Scientific 32: Ralph A. Slepecky/Visuals Unlimited 32: Dwight Kuhn 32: Kevin and Betty Collins/Visuals Unlimited 33: Dr. Susumu Ho, Harvard Medical School 35: Elizabeth Gentt/Visuals Unlimited

Chapter 5
57: D. Friend and D. Fawcett/Visuals Unlimited

Chapter 7
84: Skip Moody/Dembinsky Photo Associates 84: Dennis Drenner 84: © Paul W. Johnson/Biological Photo Service 84: J.M. Kingsbury 88: D.G. Fisher and R.F. Evert

Chapter 12
166, top: Elizabeth Gentt/Visuals Unlimited
166, bottom: David M. Phillips/Visuals Unlimited
167, left: Oliver Meckes/Ottawa/Photo Researchers, Inc.
167, right: David M. Phillips/Visuals Unlimited

Chapter 13
183: William E. Ferguson 185, top: Michael Giannechini/Photo Researchers, Inc. 185, bottom: R.J. Erwin/Photo Researchers, Inc.

Chapter 14
200: Hal Beral/Visuals Unlimited 201, top: Darrel Gulin/Dembinsky Photo Associates 201, bottom left: Dr. Paul Zahl/Photo Researchers, Inc. 201, bottom right: Alex Rakosy/Dembinsky Photo Associates 202, top: Gerald and Buff Corsi/Visuals Unlimited 202, bottom: Terry Donnelly/Dembinsky Photo Associates 203, top: Brian Parker/Tom Stack & Associates 203, center: A. Kerstitch/Visuals Unlimited 203, bottom: Dave Fleetham/Tom Stack & Associates 204, top left: William C. Jorgenson/Visuals Unlimited 204, top center: John D. Cunningham/Visuals Unlimited 204, top right: Triarch/Visuals Unlimited 204 bottom left: Stan Elems/Visuals Unlimited 204, bottom right: Wim Van Egmond 205: T.E. Adams/Peter Arnold Inc. 206, left: T.E. Adams/Visuals Unlimited 206, right: Visuals Unlimited 207, top: Andrew Syred/SPL/Photo Researchers, Inc. 207, bottom: Nuridsany et Pérennou/Photo Researchers, Inc. 208, top: John Shaw/Tom Stack & Associates 208, bottom: Parke H. John, Jr./Visuals Unlimited